INTRODUCTION TO DIGITAL FILTERING IN GEOPHYSICS

FURTHER TITLES IN THIS SERIES

1. *F.A. VENING MEINESZ*
THE EARTH'S CRUST AND MANTLE

2. *T. RIKITAKE*
ELECTROMAGNETISM AND THE EARTH'S INTERIOR

3. *D.W. COLLINSON, K.M. CREER* and *S.K. RUNCORN*
METHODS IN PALAEOMAGNETISM

4. *M. BÅTH*
MATHEMATICAL ASPECTS OF SEISMOLOGY

5. *F.D. STACEY* and *S.K. BANERJEE*
THE PHYSICAL PRINCIPLES OF ROCK MAGNETISM

6. *L. CIVETTA, P. GASPARINI, A. RAPOLLA* and *G. LUONGO (Editors)*
PHYSICAL VOLCANOLOGY

7. *M. BÅTH*
SPECTRAL ANALYSIS IN GEOPHYSICS

Developments in Solid Earth Geophysics

8

INTRODUCTION TO DIGITAL FILTERING IN GEOPHYSICS

BY

OTA KULHÁNEK

Seismological Institute
University of Uppsala, Sweden

ELSEVIER SCIENTIFIC PUBLISHING COMPANY
Amsterdam — Oxford — New York   1976

ELSEVIER SCIENTIFIC PUBLISHING COMPANY
335 Jan van Galenstraat
P.O. Box 211, Amsterdam, The Netherlands

AMERICAN ELSEVIER PUBLISHING COMPANY, INC.
52 Vanderbilt Avenue
New York, New York 10017

With 32 illustrations and 5 tables

Library of Congress Cataloging in Publication Data

Kulhánek, Ota.
 Introduction to digital filtering in geophysics.

 (Developments in solid earth geophysics ; 8)
 Bibliography: p.
 Includes index.
 1. Geophysics--Mathematical models. 2. Geophysics--Data processing. 3. Digital filters (Mathematics) I. Title. II. Series.
QE501.3.K84 551'.028'5 76-23185
ISBN 0-444-41331-6

Copyright © 1976 by Elsevier Scientific Publishing Company, Amsterdam

All rights reserved. No part of this publication may be reproduced, stored in a retrieval system, or transmitted in any form or by any means, electronic, mechanical, photocopying, recording, or otherwise, without the prior written permission of the publisher,
Elsevier Scientific Publishing Company, Jan van Galenstraat 335, Amsterdam

Printed in The Netherlands

Ota Kulhánek: Introduction to Digital Filtering in Geophysics

Corrections

Page	Line	Instead of	Read
ix	21	$\{h(t)\}$	$\mathcal{F}\{h(t)\}$
ix	22	$^{-1}\{H(\omega)\}$	$\mathcal{F}^{-1}\{H(\omega)\}$
x	14	$\{h(t)\}$	$\mathcal{L}\{h(t)\}$
x	15	$^{-1}\{H(p)\}$	$\mathcal{L}^{-1}\{H(p)\}$
xi	8	$\{x_n\}$	$\mathcal{I}\{x_n\}$
xi	9	$^{-1}\{X(z)\}$	$\mathcal{I}^{-1}\{X(z)\}$
1	19	$\{x(t)\}$	$\hat{O}\{x(t)\}$
1	20	symbol is	symbol $\hat{O}$ is
1	25	operator, , i.e.	operator, $\hat{O}$, i.e.
2	9	$\{Ax(t)\} = A \quad \{x(t)\}$	$\hat{O}\{Ax(t)\} = A\,\hat{O}\{x(t)\}$
2	10	$\left\{\sum_{i=0}^{\infty} A_i x(t)\right\} = \sum_{i=0}^{\infty}\left\{A_i x(t)\right\}$	$\hat{O}\left\{\sum_{i=0}^{\infty} A_i x(t)\right\} = \sum_{i=0}^{\infty} \hat{O}\left\{A_i x(t)\right\}$
2	30	$\{x(t)\}$	$\hat{O}\{x(t)\}$
7	4	$\{h(t)\}$	$\mathcal{F}\{h(t)\}$
7	9	$^{-1}\{H(\omega)\}$	$\mathcal{F}^{-1}\{H(\omega)\}$
9	11	$\{h(t)\}$	$\mathcal{L}\{h(t)\}$
10	4	unction	function
10	7	$\int_{-\infty}^{\infty} \int h(\eta)\, x(t+\tau-\eta)\, d\eta$	$\int_{-\infty}^{\infty} h(\eta)\, x(t+\tau-\eta)\, d\eta$
12	10	$\{x(t)\}$	$\mathcal{F}\{x(t)\}$
12	26	$A/H(\Omega)\,\|/A$	$A\,\|H(\Omega)\,\|/A$
17	9	$\sum_{n-\infty}^{\infty}$	$\sum_{n=-\infty}^{\infty}$
20	10	$\{x_n\}$	$\mathcal{I}\{x_n\}$
24	18	$\{\delta(t-n\Delta t)\}$	$\mathcal{L}\{\delta(t-n\Delta t)\}$
24	19	$\{\delta(t)\}$	$\mathcal{L}\{\delta(t)\}$
24	20	$\{x(t-n\Delta t)\}$	$\mathcal{L}\{x(t-n\Delta t)\}$
24	20	$\{x(t)\}$	$\mathcal{L}\{x(t)\}$
38	31	0	0 elsewhere
67	30	$-a_0 b_1 x_1$	$-a_1 b_1 x_1$
69	9	$z+2-z^{-1}$	$-z+2-z^{-1}$
69	19	$z+2-z^{-1}$	$-z+2-z^{-1}$
72	14	$\int_{E/2}^{E/2}$	$\int_{-E/2}^{E/2}$

Page	Line	Instead of	Read
84	8	$\Omega_1 \leq \omega < \Omega_2$	$\Omega_1 < \omega \leq \Omega_2$
85	19	$\pm \Delta\omega/\pi$	$\pm \pi/\Delta\Omega$
93	2	ϵ_A	ϵ
93	25	$\sqrt{\epsilon^2 + 1}$	$\sqrt{\epsilon^{-2} + 1}$
134	11	$v/\sin \psi$	$v/\sin \psi$
136	35	$\Phi_m - \gamma_m$	$\varphi_m - \gamma_m$
138	2	$\cos(\Phi_m - \gamma_m)$	$\cos(\varphi_m - \gamma_m)$
138	2	$\sin(\Phi_m - \gamma_m)$	$\sin(\varphi_m - \gamma_m)$
138	9	$\cos(\Phi_m - \gamma_m)$	$\cos(\varphi_m - \gamma_m)$
138	9	$\sin(\Phi_m - \gamma_m)$	$\sin(\varphi_m - \gamma_m)$
140	23	each$_M$ of	each of
140	29	s_{mi}	s_{Mi}
150	4	r_m	$\mathbf{r}_m$
150	25	$\{h_m(t-t_m)\}$	$\mathcal{F}\{h_m(t-t_m)\}$
150	25	$H_m(f,k)$	$H_m(f,\mathbf{k})$
150	25	$\exp(-j2\pi r_m \cdot k)$	$\exp(-j2\pi \mathbf{r}_m \cdot \mathbf{k})$
150	30	$\exp(-j2\pi r_m \cdot k)$	$\exp(-j2\pi \mathbf{r}_m \cdot \mathbf{k})$

PREFACE

Modern measuring techniques in any technical or scientific field, geophysics included, do not provide noiseless observational data. Therefore, it is very likely that when starting to evaluate their observations students and young researchers sooner or later will meet up with the signal–noise separation or the filtering problem in all its complexity. The theory of filters and related problems has been already adequately treated in literature on communication systems and elsewhere. Nevertheless, for students in seismology and other geophysical disciplines to read textbooks on communication problems and to orient themselves in the extensive literature may be rather burdensome.

An attempt is made here to present available information pertinent to filtering techniques employed in geophysics in a systematic, tutorial form. Much of the collected information is not new, but is scattered rather widely in the geophysical literature and literature on communication systems. Emphasis is laid on understandable presentation of principles of various filtering techniques for the benefit of students and junior researchers in geophysics. Applications are covered by references mostly dealing with seismological problems, but other geophysical branches like gravity, geomagnetism, etc. are also represented. Although the references are numerous and include items up to the beginning of 1974, they are by no means meant to be complete. Rather, typical examples have been selected according to the author's opinion. It has been assumed that the filtering operation itself is simulated by digital computers; however, no attempt has been made to include computer programs. These may be found elsewhere.

Since data filtering is a problem common to the analysis of any recorded physical quantity, young researchers from other than geophysical fields may also find the text helpful. The presentation is suitable for readers whose mathematical background includes complex variable and integral transforms. Chapters 1 and 2 present the principles of digital processing and basic approaches in the filter design. Chapters 3 through 6 deal in more detail with filter categories which are likely to be utilized in solving various geophysical problems.

I was originally motivated to write this book by Professor Markus Båth with whom I share the feeling of a need for an introductory textbook on filtering techniques in geophysics. The selection of topics and the presentation used are influenced by lectures I have been giving on the subject at the Seismological Institute, Uppsala.

It is my pleasure to take this opportunity to thank Professor Markus Båth for his significant assistance in preparing this book, especially for critically reading the manuscript and for valuable comments and suggestions improving the text. My thanks are also due to my colleague Dr. John S. Farnbach for useful discussions and for assistance in checking the language. The facilities of the Seismological Institute were put at my disposal for preparation of the manuscript. Gratitude is expressed to Miss Ulla Hjelmqvist for typing a part of the final manuscript and to Mrs. Ester Dreimanis for drafting the figures. The financial support provided by the Swedish Natural Science Research Council is gratefully acknowledged. Finally, I like to thank my wife Iva for typing the rough draft as well as a part of the final manuscript. Her encouragement and continuous optimism contributed considerably to this book in various phases of its evolution.

Seismological Institute
Uppsala, Sweden

OTA KULHÁNEK
January, 1975

CONTENTS

 PAGE

PREFACE.. V

LIST OF SYMBOLS... IX

CHAPTER 1. FUNDAMENTAL RELATIONS.. 1

  1.1 Definitions.. 1
  1.2 Relationships between input and output signals........................ 3
      1.2.1 Relationships in time domain................................... 3
      1.2.2 Relationships in frequency domain.............................. 6
  1.3 Relationships between input and output statistical characteristics..... 9
  1.4 Signal distortion due to the transmission through a filter............ 12
  1.5 Analog–digital conversion... 15
  1.6 z-transform... 20
  1.7 Input–output relationships for digital systems........................ 27
  1.8 Stability of digital systems.. 33

CHAPTER 2. DESIGN PRINCIPLES OF DIGITAL FILTERS......................... 35

  2.1 Nonrecursive filtering.. 35
  2.2 Recursive filtering... 40
      2.2.1 First-order recursive filtering................................ 42
      2.2.2 Second-order recursive filtering............................... 44
      2.2.3 Lth-order recursive filtering................................ 46
  2.3 System functions of recursive filters................................. 48
  2.4 Pole-zero technique... 50
  2.5 Approximation of analog systems....................................... 55
      2.5.1 Impulse invariance... 55
      2.5.2 Convolution approximation...................................... 59
      2.5.3 Approximation by means of bilinear transformation.............. 61
  2.6 Phase-distortionless filters.. 65
  2.7 Effects of quantization in digital filters............................ 71
      2.7.1 Errors caused by input quantization............................ 71
      2.7.2 Errors caused by product quantization.......................... 72
      2.7.3 Errors caused by filter-parameter quantization................. 76

CHAPTER 3. LOW-, HIGH- AND BAND-PASS FILTERS............................ 77

  3.1 Low-pass filters.. 78
      3.1.1 Ideal low-pass filters... 78
      3.1.2 Truncated unit-impulse response function....................... 81
      3.1.3 Ormsby and Martin-Graham filters............................... 82
      3.1.4 Butterworth filters.. 87
      3.1.5 Chebyshev filters.. 92
  3.2 High-pass filters... 94
  3.3 Band-pass filters... 96

CHAPTER 4. CORRELATION AND OPTIMUM FILTERS 99

4.1 Correlation filters 99
    4.1.1 Detection of periodic signals by autocorrelation 99
    4.1.2 Detection of signals by cross-correlation 101
4.2 Matched filters 105
4.3 Wiener optimum filters 108
    4.3.1 Optimum noncausal systems 109
    4.3.2 Considerations in designing Wiener optimum filters 112
    4.3.3 Multi-dimensional Wiener optimum filters 116
4.4 Polarization filters 116

CHAPTER 5. DECONVOLUTION FILTERS 120

5.1 Exact deconvolution 120
5.2 Truncated approximate deconvolution 122
5.3 Least-squares approximate deconvolution 124
5.4 Trace decomposition 127

CHAPTER 6. MULTI-DIMENSIONAL FILTERS 130

6.1 SS technique 131
6.2 Basic requirements for array patterns 135
6.3 DS and WDS techniques 136
6.4 Improvement of the signal/noise ratio due to DS and WDS techniques 138
6.5 Velocity filtering 142
    6.5.1 Signal and noise characteristics 142
    6.5.2 Velocity filtering by means of one-dimensional arrays 143
    6.5.3 Two-dimensional wavenumber filtering 148
    6.5.4 Velocity filtering by means of two-dimensional arrays 149

REFERENCES 154

SUBJECT INDEX 163

LIST OF SYMBOLS

a_i, b_i	ith coefficient in the numerator and denominator of $H(z)$, respectively
A, K, c, d	real constants
A_i, B_i	ith zero and pole, in the z-plane, of $H(z)$, respectively
cps	cycles per second
$C_N(\omega)$	Chebyshev polynomial of degree N
d	unit-impulse response sequence of deconvolution filter
d_n	nth sample of d
dB	decibel
$D(z)$	system function of a deconvolution filter
e	base of natural logarithm system
e	error sequence
e_n	nth sample of e
e_{in}	nth sample of e_i; index i denotes the approximation used
$e(t), e(\omega), e(\omega, N)$	error functions
f	cyclic variable frequency
f_s	sampling cyclic frequency
f_N	Nyquist cyclic frequency (folding cyclic frequency)
$\{h(t)\}$	Fourier transform of $h(t)$
$^{-1}\{H(\omega)\}$	inverse Fourier transform of $H(\omega)$
g, y	sequences of output samples
g_n, y_n	nth sample of g and y, respectively
g_{in}	nth sample of g_i; index i denotes the approximation used
$g(t), y(t)$	analog output functions
$G(z), Y(z)$	z-transform of g and y, respectively
$h, \{h_n\}$	unit-impulse response sequence
$\hat{h}$	truncated h
h_n	nth sample of h
$h(t)$	analog unit-impulse response function (analog weighting function)
$H(p)$	transfer function
$H(z)$	system function (pulse-transfer function)
$H(\omega)$	frequency response function (response function, system function)

LIST OF SYMBOLS

$H^*(\omega)$, $X^*(\omega)$	complex conjugate of $H(\omega)$ and $X(\omega)$, respectively
$H_B(p)$, $H_C(p)$	transfer function of a Butterworth and Chebyshev filter, respectively
$H_L(\omega)$ $H_H(\omega)$	frequency response function of a low- and high-pass filter, respectively
$i, k, l, m, n, L, M, N, P, R$	integers
j	$\sqrt{-1}$
k, k_x, k_y	wavenumbers
$\mathbf{k}$	vector wavenumber
k_N	highest resolvable wavenumber
ln	natural logarithm
log	logarithm to the base 10
$\mathcal{L}\{h(t)\}$	Laplace transform of $h(t)$
$\mathcal{L}^{-1}\{H(p)\}$	inverse Laplace transform of $H(p)$
M	constant gain factor
$M(\omega)$	amplitude response (gain factor)
$M_B(\omega)$, $M_C(\omega)$	amplitude response of a Butterworth and Chebyshev filter, respectively
$M_L(\omega)$, $M_H(\omega)$	amplitude response of a low- and high-pass filter, respectively
n_i	sequence of noise samples in the ith channel
n_{in}	nth sample of n_i
$n(t)$	analog noise
p	complex variable
p_i	ith pole of $H(p)$
p_i^*	complex conjugate of p_i
$P(\omega)$, $Q(\omega)$	real and imaginary part of $H(\omega)$, respectively
Q	signal/noise ratio
$R_{xx}(\tau)$	autocorrelation of $x(t)$
Re, Im	real and imaginary part, respectively
s_i	sequence of signal samples in the ith channel
s_{in}	nth sample of s_i
$s(t)$	analog signal
t, t_1, τ, τ_1	time
V	apparent velocity
w	time-window sequence
w_n	nth sample of w
$w(t)$	time window
$W(\omega)$	Fourier transform of $w(t)$
$x, \{x_n\}$	sequence of input samples
x_n	nth sample of x
$x_{\hat{S}}(t)$	output of the sampler

LIST OF SYMBOLS

$x(t)$	analog input
$X(\omega)$	Fourier transform of $x(t)$
$y(k, \theta)$	array response function
$y_d(t)$	desired analog output
z	complex variable
z_i	ith pole of $H(z)$
$\{x_n\}$, $X(z)$	z-transform of x
$^{-1}\{X(z)\}$	inverse z-transform of $X(z)$
α_i, β_i	ith zero and pole, in the z^{-1} plane, of $H(z)$, respectively
$\delta, \{\delta_n\}$	unit-impulse sequence
$\delta(t)$	Dirac delta function
$\delta\Delta_t(t)$	ideal sampling function
Δt	sampling period
Δx	sensor spacing
λ	wavelength
σ	real part of p
$\phi(\omega)$	phase response
$\Phi_{xx}(\omega)$	power spectral density of $x(t)$
ω	angular variable frequency
$\omega_c, \omega_L, \omega_H$	angular cut-off frequencies
ω_s	sampling angular frequency
ω_N	Nyquist angular frequency (folding angular frequency)
Ω	constant angular frequency

Chapter 1

FUNDAMENTAL RELATIONS

The term *filter* is used to characterize a system which can perform an effective prescribed separation of the desired information carried by the signal from the unwanted portion commonly called *noise*. The separation may be carried out e.g. on the basis of frequency, velocity, or polarization discrimination by *frequency, velocity*, or *polarization filters*, respectively. Throughout the present book it will be assumed that the filter characteristics of interest will be simulated on digital computers. Therefore we put all the emphasis on the investigation of input–output relations (also called *terminal relations*), in other words on relations between the filter excitation and corresponding response. Consequently, the internal configuration of a physical system, i.e. filter design by means of electronic, mechanical, optical and other components is not discussed.

1.1 DEFINITIONS

When digital computer simulations are used, the word *operator* may rather well be substituted for "filter". In general, our effort here will be to specify a transformation of the input function, $x(t)$, into the corresponding output function, $y(t)$, so that:

$$y(t) = \{x(t)\} \qquad [1.1]$$

where the symbol is the operator describing the transformation. The input and output in [1.1] are considered to be functions of time. In Chapter 6 we shall discuss filters where the input and output are functions of other variables such as time and space. Throughout the present book only real-valued input and output functions will be considered.

A design of the operator, , i.e. the determination of its mathematical form based on the previous knowledge and experience and respecting various a-priori filtering requirements, is called *filter synthesis*. An inverse procedure, *filter analysis*, means an investigation of properties of a given operator. Broadly speaking, the filter synthesis is a complicated process, since [1.1] should provide an exact response to any applied input. However, for certain categories of filters this process is significantly simplified. In order to simplify the introduction into the filter synthesis, we shall first limit ourselves to single-channel filters which are linear, time invariant, stable and causal.

Filters with one input and one output function are called *single-channel filters*. All other filters are *multi-channel filters* covering *multiple-input, multiple-output* and *multiple-input–multiple-output systems*.

A system is called *linear* if the proportionality and superposition laws hold for any input function $x(t)$ and arbitrarily chosen constant A. That is:

$$\{Ax(t)\} = A\{x(t)\} \qquad \text{proportionality law or homogeneous property}$$

$$\left\{\sum_{i=0}^{\infty} A_i x(t)\right\} = \sum_{i=0}^{\infty} \{A_i x(t)\} \qquad \text{superposition law or additive property} \qquad [1.2]$$

An operator which satisfies condition [1.2] is called *linear operator*.

If the properties of the system, influencing the input–output relation, do not vary with time the system is said to be *time-invariant* or with *constant parameters* or *fixed parameters*. Time invariance means that if the system were described by a differential equation involving the input and output, the coefficients of the equation would be time-independent. Strictly speaking, all physical filters do change their characteristics with time due to the variations of temperature, pressure, etc. and are therefore *time-varying filters*. However, these variations are very slow when compared with the duration of the input signals processed and therefore, if not specified otherwise, they will be neglected throughout this book.

Stability of a system becomes an important topic especially when the output signal, or a certain part of it, is recirculated back into the input (so-called *feedback systems*). A system is considered to be *stable* when every bounded input produces a bounded output. This condition can be expressed in the following way: if $|x(t)| \leq A$, then $|y(t)| \leq KA$ where A and K are finite constants which are independent of the input. The stability problem is discussed in more detail in the next section and especially in Section 1.8.

Due to the basic principles of cause and effect the output of any physical system cannot precede the corresponding input in time. That is, if $x(t) = 0$ for $t < t_1$ then:

$$y(t) = \{x(t)\} = 0 \quad \text{for } t < t_1 \qquad [1.3]$$

The condition [1.3] is called the *causality condition*. A system or an operator satisfying this condition is said to be *causal* or *physically realizable*. While for physical filters causality is a decisive requirement, in digital-computer simulations it loses most of its importance.

From these five fundamental definitions one may feel that single-channel, linear, stable, time-invariant and causal filters represent a small and exclusive group of filters. However, as will be shown later, the contrary is true. These filters (for computer-simulated filters causality property excluded) form the most common category.

1.2 RELATIONSHIPS BETWEEN INPUT AND OUTPUT SIGNALS

We shall now derive the relationships between the excitation and response of a filter in time as well as in frequency domain, assuming that the filter is linear, causal and time-invariant. In the most general case of filter synthesis, the input signal is given along with further requirements describing the form of the so-called *desired output*. Then, the goal of the synthesis is to design the characteristics of the filter in a way which would guarantee reasonable resemblance between the true output and the desired output. Relationships between the three quantities, the input, output and system characteristics, are essential for any filter design.

1.2.1 *Relationships in time domain*

Consider that at time $t = 0$ a *unit impulse* $\delta(t)$ is applied at the input. The unit impulse or *Dirac delta function* can be visualized as a rectangular pulse of an infinitesimal width dt and unit area beneath the pulse:

$$\int_{-\infty}^{\infty} \delta(t) \, dt = 1$$

Outside the time interval dt the amplitude of the pulse is zero (Papoulis, 1962, p. 270; Båth, 1968, p. 247). We denote the response of the system to the unit impulse by $h(t)$. The function $h(t)$ is frequently called the *unit-impulse response*. The system is initially at rest, which will be always the case unless otherwise stated. Imposing further the causal property, it is of necessity that the unit-impulse response $h(t) = 0$ for $t < 0$. For a filter with constant parameters an application of the unit impulse delayed by time $t = \tau$ will result in a response delayed by the same time interval. So, for the unit pulse $\delta(t - \tau)$ at the input the filter produces output $h(t - \tau)$. If the area of the input pulse is not equal to 1 but equal to an arbitrary constant A, then the response to a delayed input pulse would be $Ah(t - \tau)$, provided that the linearity property holds.

The unit impulse as an input function is of course a special case. Consider therefore an arbitrary time function $x(t)$ acting as the filter input. The input function can be considered to be made up of a large number of rectangular pulses, each of them of infinitesimal width dt. Each of these rectangles, when passing through the filter, acts like a separate input pulse and produces its own output. The output resulting from the passage of the whole input $x(t)$ through filter, can be considered as a sum of all these separate outputs, respecting the proper time relations. For example, at time $t = \tau$ the input pulse has the width $d\tau$ and the height $x(\tau)$ and its area is $x(\tau)d\tau$. It follows from the discussion above that the corresponding differential of the output is:

$$dy(t) = x(\tau) \, d\tau \, h(t - \tau) \qquad [1.4]$$

Let us suppose that the input function has been substituted by narrow rectangles in its entire history, starting from the infinite past and continuing up to the time t. Then, summing all the output differentials, $dy(t)$ in [1.4], due to the separate input pulses, the output at time t may be expressed as:

$$y(t) = \int_{-\infty}^{t} x(\tau) h(t-\tau) d\tau \qquad [1.5]$$

The right-hand side integral, called the *convolution* or *superposition integral*, shows that a system can be characterized (in terms of the cause and the effect) by an integral operator. The input–output relationship of any linear, time-invariant system is given by the function $h(t)$. Equation [1.5] provides the relation between the input, output and unit-impulse response and is therefore of the utmost importance in filter synthesis. Since the three functions involved are functions of time, synthesis which makes use of [1.5] is called the *time-domain synthesis*.

The principles of physical (analog) and digital convolution processes are further discussed by Dobrin and Ward (1962), Anstey (1964) and Silverman (1967) among others. An interesting explanation of the superposition integral by means of a delay-line apparatus has been presented by Jones et al. (1955).

The form of $h(t)$ is decisive in estimating the extent to which the past of the input $x(t)$ before the instant t influences the output at the instant t. By the length of the response we mean the time interval during which $h(t)$ is significantly different from zero. Generally speaking, the longer the unit-impulse response, the longer the history of the input which contributes to the output value at time t.

Equation [1.5] may be modified and various forms of the convolution integral may be obtained. For example, referring to the causal property, we note that since $h(t-\tau) = 0$ for $\tau > t$ there will be no contribution to the integral value in [1.5] from the interval $t < \tau < \infty$. So, we may replace the upper limit of the integration by infinity and rewrite the formula in the following way:

$$y(t) = \int_{-\infty}^{\infty} x(\tau) h(t-\tau) d\tau \qquad [1.6]$$

As follows from [1.5] and [1.6] the process of convolution consists of four operations: displacement, time reversal or folding, multiplication and integration. Suppose that two time functions $x(t)$ and $h(t)$ are the input and the unit-impulse response of a filter, respectively. The corresponding output at time $t = t_1$ may be obtained by applying successively the four operations. First we displace the response by t_1, then we fold the displaced function about the axis $t = 0$ and multiply it by $x(t)$. The output for $t = t_1$ is then the integral of this product and the integration is carried out along the entire range of the variable t. Table I shows the four operations.

Another form of the convolution integral may be obtained by using a single substitution $\tau_1 = t - \tau$. It follows that:

INPUT AND OUTPUT SIGNALS—RELATIONSHIPS

$$y(t) = \int_{-\infty}^{\infty} h(\tau_1) \, x(t - \tau_1) d\tau_1 \qquad [1.7]$$

TABLE I

Four convolution operations

Signals to be convolved	$x(t)$	$h(t)$
1. Time shift by t_1		$h(t + t_1)$
2. Folding about $t = 0$		$h(t_1 - t)$
3. Multiplication	$x(t) h(t_1 - t)$	
4. Integration	$y(t_1) = \int_{-\infty}^{\infty} x(t) h(t_1 - t) \, dt$	

The output $y(t)$ may be viewed as a weighted sum over the whole history of the input $x(t)$. Weights are given by the unit-impulse response, $h(t)$, which is sometimes also called the *weighting function*. Equations [1.6] and [1.7] prove the important property that the result of convolution is the same regardless of which of the two convolved functions is displaced if the displaced function is also the one folded. In a linear filter the impulse response and input are interchangeable. Assuming the causal property, the integration in [1.7] can be started at $\tau_1 = 0$ since there is no contribution to the integral value for $\tau_1 < 0$. Hence, the output may be written as:

$$y(t) = \int_{0}^{\infty} h(\tau_1) \, x(t - \tau_1) d\tau_1 \qquad [1.8]$$

Equations [1.5]–[1.8] show various possibilities for expressing the output $y(t)$ in terms of $x(t)$ and $h(t)$, in the time domain. In practical applications some of the notations may be more useful than the others. In fact, in deriving the input–output relations, responses to types of functions other than the unit impulse may be considered as well. For example, relations which make use of a *unit-step response* may be presented (see e.g. Papoulis, 1962, pp. 87–88). These relations are especially important for technical applications, since unit-step functions are easier to generate than unit-impulse functions in a practical system.

The relationship between the three functions $x(t)$, $y(t)$ and $h(t)$, as given in [1.8], describes the filtering process (i.e. the passage of a signal through a filter) in terms of the known input signal and the filter characteristics. Two more cases may be considered. Firstly, an inverse process is to derive the input from the known output and filter characteristics. This process, called *signal restoration*, is utilized in seismological applications where it is sometimes desired to eliminate the influence of the recording system in order to obtain the true ground motion (see Chapter 5). Here the seismogram is considered to be the output of the system and the seismometer–galvanometer coupling represents the filter. A second case which may arise is to find the system response function from given input and output signals. This is the so-called *system identification* problem.

The stability of the system may also be tested by using the unit-impulse response. Consider a linear, time-invariant system with a bounded input. Then there exists a constant A such that:

$$|x(t)| \leq A \quad \text{for all } t$$

According to [1.7] the absolute output value is:

$$|y(t)| = \left| \int_{-\infty}^{\infty} h(\tau) x(t-\tau) \, d\tau \right| \leq \int_{-\infty}^{\infty} |h(\tau)| \, |x(t-\tau)| \, d\tau$$

From the bounded input condition it follows that:

$$|y(t)| \leq A \int_{-\infty}^{\infty} |h(\tau)| \, d\tau$$

Supposing that the impulse response is absolutely integrable, we write:

$$K = \int_{-\infty}^{\infty} |h(\tau)| \, d\tau < \infty \qquad [1.9]$$

then the output is bounded as well since:

$$|y(t)| \leq KA$$

and the system is stable. Hence, the absolute integrability of the unit-impulse response assures the stability of the system.

Time-varying systems cannot be characterized by a single unit-impulse response since the system parameters are time-dependent. Instead a family of functions, i.e. time-variable weighting $h(t_1, t)$, has to be applied. Function $h(t_1, t)$ is the response of the system measured at time t caused by an impulse applied at time $t - t_1$. The difference $t - t_1$ determines the time interval between the input impulse and the output observation. Clarke (1968) calls $t - t_1$ the "age" of the output. Replacing $h(\tau_1)$ in [1.8] by the more general $h(t_1, t)$ the output of a time-varying linear system takes the form:

$$y(t) = \int_{-\infty}^{\infty} h(t_1, t) x(t - t_1) \, dt_1 \qquad [1.10]$$

It follows from [1.10] that when calculating the output of systems characterized by a two-dimensional weighting function not only the time interval $t - t_1$ is important but also the particular time t at which the system is being observed. These systems are physically realizable when $h(t_1, t) = 0$ for $t_1 < 0$, independent of t.

1.2.2 Relationships in frequency domain

A linear, time-invariant system can also be characterized by its *frequency response*

INPUT AND OUTPUT SIGNALS—RELATIONSHIPS

function $H(\omega)$ (usually abbreviated to *response function*) or *system function* which is defined as the *Fourier transform* of the unit-impulse response $h(t)$:

$$H(\omega) = \mathscr{F}\{h(t)\} = \int_{-\infty}^{\infty} h(t)\, e^{-j\omega t}\, dt \qquad [1.11]$$

where the real variable ω is the angular frequency. It is noteworthy that the function $H(\omega)$ is here merely a function of angular frequency. For non-linear systems $H(\omega)$ depends also on the input applied and for time-varying systems it is also a function of time. The inverse Fourier transform gives $h(t)$ from $H(\omega)$:

$$h(t) = \mathscr{F}^{-1}\{H(\omega)\} = \frac{1}{2\pi} \int_{-\infty}^{\infty} H(\omega)\, e^{j\omega t}\, d\omega \qquad [1.12]$$

The frequency response, $H(\omega)$, is a complex function of a real variable and may be written in the form:

$$H(\omega) = P(\omega) + jQ(\omega) = M(\omega)\, e^{j\phi(\omega)} \qquad [1.13]$$

where $M(\omega) = |H(\omega)| = [P^2(\omega) + Q^2(\omega)]^{1/2}$ is the *amplitude response* or the *gain factor* and $\phi(\omega) = \arctan[Q(\omega)/P(\omega)]$ is the *phase response* of the system. The amplitude response gives the gain for a component of frequency ω in the signal passing through the system. The slope of the phase response for any frequency determines the time delay caused by the filter for that particular frequency.

The two characteristics $M(\omega)$ and $\phi(\omega)$ together completely describe the filter in terms of signal frequencies which might pass through it. By dividing $h(t)$ into its even and odd components (see e.g. Papoulis, 1962, pp. 12–13) it is possible to derive the following symmetry properties of systems with real-valued inputs and outputs:

$$|H(\omega)| = |H(-\omega)| \quad \text{and} \quad \phi(-\omega) = -\phi(\omega)$$

Letting $X(\omega)$ and $Y(\omega)$ be the complex spectra of the input and output, respectively, and applying the convolution theorem to [1.6], we have a corresponding relation in the *frequency domain:*

$$Y(\omega) = X(\omega)\, H(\omega) = X(\omega)\, M(\omega)\, e^{j\phi(\omega)} \qquad [1.14]$$

Equation [1.14] is the frequency domain equivalent to [1.6] and relates the three functions in terms of their spectra. Thus, if we apply the inverse Fourier transform, [1.12], to the right-hand product in [1.14] we obtain the output as a function of time:

$$y(t) = \frac{1}{2\pi} \int_{-\infty}^{\infty} X(\omega)\, H(\omega)\, e^{j\omega t}\, d\omega = \frac{1}{2\pi} \int_{-\infty}^{\infty} X(\omega)\, M(\omega)\, e^{j[\omega t + \phi(\omega)]}\, d\omega \qquad [1.15]$$

Comparing [1.6] and [1.15] we find that a linear, time-invariant system is equally

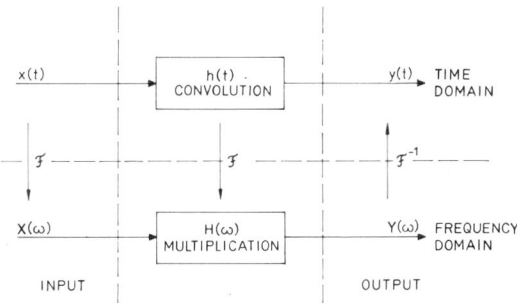

Fig. 1.1. Principal description of procedures in the time and frequency domain. Symbols $\mathcal{F}$ and $\mathcal{F}^{-1}$ are used to designate Fourier and inverse Fourier transform, respectively.

well described by its unit-impulse response or frequency response function by using the time- or frequency-domain approach, respectively. In the time domain the passage of a signal through the filter may be calculated directly by using the convolution integral, while in the frequency domain the Fourier and inverse Fourier transforms are used. Figure 1.1 shows the principles of the procedure in both domains.

Further as follows from [1.13] the system function is also defined by its real and imaginary part or by its amplitude and phase response. Filter design, in the frequency domain, via amplitude and phase characteristics is the most common approach. This is due to the fact that the desired filter properties are usually more easily expressed in terms of its amplitude and phase response. Time-domain design, via the unit-impulse response, is less frequently used. Frequency functions $P(\omega)$, $Q(\omega)$ have more or less only theoretical significance.

For large categories of system functions $M(\omega)$ and $\phi(\omega)$ as well as $P(\omega)$ and $Q(\omega)$ are not independent of each other but one of them can be uniquely determined from the other by making use of the *Hilbert transform*. Discussion of this problem goes beyond the scope of this book and will be omitted here. Interested readers are referred e.g. to Lee (1960), Papoulis (1962) or Bracewell (1965).

Another concept for describing system behaviour in the frequency domain is to make use of the *transfer function* $H(p)$ which is defined as the *Laplace transform* of the unit-impulse response:

$$H(p) = \mathcal{L}\{h(t)\} = \int_0^\infty h(t) e^{-pt} \, dt \quad \text{for } p = \sigma + j\omega \qquad [1.16]$$

Here we suppose that function $h(t)$ is defined for every $t \geq 0$, so-called *unilateral* or *one-sided Laplace transform* (hereafter, just Laplace transform). The corresponding version which considers functions defined in the entire interval of the real variable t is called the *bilateral* or *two-sided Laplace transform*. Note that while the Fourier integral $H(\omega)$ is treated as a function of the real angular frequency ω, the Laplace transform $H(p)$ is a function of the complex variable $p = \sigma + j\omega$. To

INPUT AND OUTPUT—STATISTICAL CHARACTERISTICS

assure the necessary convergence in [1.16] for the type of time functions which frequently appear in geophysical observations it is usually sufficient that the real part of p is nonnegative, i.e. $\mathrm{Re}(p) \geq 0$. It follows from the integrals in [1.11] and [1.16] that when the real part of p is zero, i.e. $\sigma = 0$, the Fourier and Laplace transforms are equal for all causal systems.

Laplace transform provides a useful tool for investigating system stability in terms of the transfer functions. The linear time-invariant system is stable if $H(p)$ has no poles in the right half of the p-plane, including the $j\omega$-axis. This stability criterion follows from Laplace transform relationships. A weighting function $h(t)$ which tends to zero sufficiently rapidly for $t \to \infty$ has a transform $\{h(t)\}$ with all poles located to the left of the imaginary axis.

The investigation of causal properties of the system by means of the transfer function is rather complicated and no general solution to this problem exists (see Papoulis, 1962, pp. 212–214).

1.3 RELATIONSHIPS BETWEEN INPUT AND OUTPUT STATISTICAL CHARACTERISTICS

Besides the convolution integral defining the relationship between the input, $x(t)$, and the output, $y(t)$, of a linear system, close dependences of great importance may be found also between the corresponding autocorrelation functions $R_{xx}(\tau)$ and $R_{yy}(\tau)$. As in the case of the convolution, the autocorrelation relationship also has an equivalent in the frequency domain which can be described in terms of power-spectral densities (hereafter just power spectra) $\Phi_{xx}(\omega)$ and $\Phi_{yy}(\omega)$. Formulae are also known which make use of the input–output cross-correlation function $R_{xy}(\tau)$. The utilization of correlation functions and/or power spectra is advantageous especially in noise studies and when signals of random nature are to be processed. This section briefly reviews some of the essential dependences. More detailed explanation may be found in Wadsworth et al. (1953), Lee (1960) or Solodovnikov (1960). It will be assumed that the system is linear and time-invariant and that the input and output are *stationary random signals* (i.e. signals whose statistical properties do not vary over time). Consider a random output signal resulting from application of the random input $x(t)$:

$$y(t) = \int_{-\infty}^{\infty} h(\lambda)\, x(t - \lambda)\, \mathrm{d}\lambda$$

and the time-delayed signal:

$$y(t + \tau) = \int_{-\infty}^{\infty} h(\eta)\, x(t + \tau - \eta)\, \mathrm{d}\eta$$

where both the variables λ and η have a physical meaning of time and $h(t)$ is the unit-impulse response of the system. Let us derive the form of the autocorrelation function $R_{yy}(\tau)$ and its relation to $R_{xx}(\tau)$.

According to the definition (Lee, 1960, p. 51) the *autocorrelation unction* of the stationary output signal is:

$$R_{yy}(\tau) = \lim_{T_1 \to \infty} \frac{1}{2T_1} \int_{-T_1}^{T_1} y(t)\, y(t+\tau)\, dt$$

$$= \lim_{T_1 \to \infty} \frac{1}{2T_1} \int_{-T_1}^{T_1} dt \int_{-\infty}^{\infty} h(\lambda)\, x(t-\lambda)\, d\lambda \int_{-\infty}^{\infty} h(\eta)\, x(t+\tau-\eta)\, d\eta$$

By changing the order of the limiting operation and integrations we find:

$$R_{yy}(\tau) = \int_{-\infty}^{\infty} h(\lambda)\, d\lambda \int_{-\infty}^{\infty} h(\eta) \left[\lim_{T_1 \to \infty} \frac{1}{2T_1} \int_{-T_1}^{T_1} x(t-\lambda)\, x(t+\tau-\eta)\, dt \right] d\eta$$

Since the quantity in brackets is the input autocorrelation, $R_{xx}(\tau+\lambda-\eta)$, the output autocorrelation function becomes:

$$R_{yy}(\tau) = \int_{-\infty}^{\infty} h(\lambda)\, d\lambda \int_{-\infty}^{\infty} h(\eta)\, d\eta\, R_{xx}(\tau+\lambda-\eta) \qquad [1.17]$$

By letting τ be zero we receive the mean-square value of the random output signal as:

$$\overline{y^2(t)} = R_{yy}(0) = \int_{-\infty}^{\infty} h(\lambda)\, d\lambda \int_{-\infty}^{\infty} h(\eta)\, d\eta\, R_{xx}(\lambda-\eta) \qquad [1.18]$$

For certain applications another form of [1.17] may be more suitable. In order to simplify the relation we introduce a substitution $\nu = \eta - \lambda$, which yields:

$$R_{yy}(\tau) = \int_{-\infty}^{\infty} h(\lambda)\, d\lambda \int_{-\infty}^{\infty} h(\nu+\lambda)\, d\nu\, R_{xx}(\tau-\nu)$$

Rearranging the order of integration we find:

$$R_{yy}(\tau) = \int_{-\infty}^{\infty} R_{xx}(\tau-\nu) \left[\int_{-\infty}^{\infty} h(\lambda)\, h(\lambda+\nu)\, d\lambda \right] d\nu$$

where, due to the definition of the autocorrelation of a transient function (Lee, 1960, p. 37), the expression in the brackets is the autocorrelation of the unit-impulse response, $R_{hh}(\tau)$. Thus:

$$R_{yy}(\tau) = \int_{-\infty}^{\infty} R_{hh}(\nu)\, R_{xx}(\tau-\nu)\, d\nu \qquad [1.19]$$

which is a relation analogous to [1.6] for time functions $x(t)$, $y(t)$ and $h(t)$. The output mean-square value may alternatively be expressed as:

INPUT AND OUTPUT—STATISTICAL CHARACTERISTICS

$$\overline{y^2(t)} = R_{yy}(0) = \int_{-\infty}^{\infty} R_{hh}(\nu) R_{xx}(\nu) \, d\nu \qquad [1.20]$$

where the evenness of any autocorrelation function has been exploited.

Equation [1.19] presents an interesting result: the output autocorrelation is determined by the impulse response and input autocorrelation without the knowledge of the output itself.

Another important relation for processing random signals is that involving the input–output *cross-correlation function*, $R_{xy}(\tau)$. According to the definition:

$$R_{xy}(\tau) = \lim_{T_1 \to \infty} \frac{1}{2T_1} \int_{-T_1}^{T_1} x(t) y(t+\tau) \, dt$$

Since:

$$y(t+\tau) = \int_{-\infty}^{\infty} h(\eta) x(t+\tau-\eta) \, d\eta$$

the input–output cross-correlation may be written after inverting the order of the integrations and the limiting operation:

$$R_{xy}(\tau) = \int_{-\infty}^{\infty} h(\eta) \left[\lim_{T_1 \to \infty} \frac{1}{2T_1} \int_{-T_1}^{T_1} x(t) x(t+\tau-\eta) \, dt \right] d\eta$$

where the quantity in brackets is the input autocorrelation $R_{xx}(\tau-\eta)$. Thus, $R_{xy}(\tau)$ becomes:

$$R_{xy}(\tau) = \int_{-\infty}^{\infty} h(\eta) R_{xx}(\tau-\eta) \, d\eta \qquad [1.21]$$

The input–output cross-correlation is the convolution of the unit-impulse response and the input autocorrelation. In other words, when the linear system is excited by the function $R_{xx}(\tau)$ it will respond with $R_{xy}(\tau)$.

The autocorrelation function and the corresponding power spectrum form a Fourier transform pair. Therefore, the output *power spectrum* is:

$$\Phi_{yy}(\omega) = \int_{-\infty}^{\infty} R_{yy}(\tau) e^{-j\omega\tau} \, d\tau \qquad [1.22]$$

Introducing the expression for $R_{yy}(\tau)$ from [1.17] and terms $e^{\pm j\omega\lambda}$ and $e^{\pm j\omega\eta}$ and separation of variables yield:

$$\Phi_{yy}(\omega) = \int_{-\infty}^{\infty} h(\eta) e^{-j\omega\eta} \, d\eta \int_{-\infty}^{\infty} h(\lambda) e^{j\omega\lambda} \, d\lambda \int_{-\infty}^{\infty} R_{xx}(\tau+\lambda-\eta) e^{-j\omega(\tau+\lambda-\eta)} \, d\tau$$

where, according to [1.11], the first integral is the frequency response function, $H(\omega)$, of the system. The second integral is the complex conjugate of the response function, $H^*(\omega)$, and finally due to the definition [1.22], the third integral is the input power spectrum $\Phi_{xx}(\omega)$. Thus, the power spectrum of the output becomes:

$$\Phi_{yy}(\omega) = H(\omega)\, H^*(\omega)\, \Phi_{xx}(\omega) = |H(\omega)|^2\, \Phi_{xx}(\omega) \qquad [1.23]$$

1.4 SIGNAL DISTORTION DUE TO THE TRANSMISSION THROUGH A FILTER

Suppose a simple harmonic signal of a form:

$$x(t) = A \sin \Omega t \qquad A, \Omega = \text{constant}$$

which has been applied for an infinitely long time to the input (steady-state signal) of a linear time-invariant system. The corresponding Fourier transform is:

$$X(\omega) = \mathscr{F}\{x(t)\} = (1/j)\, A\pi\, [\delta(\omega - \Omega) - \delta(\omega + \Omega)]$$

According to [1.15] the response of the system to the applied input $x(t)$ will be:

$$y(t) = \frac{1}{2\pi} \int_{-\infty}^{\infty} X(\omega)\, H(\omega)\, e^{j\omega t}\, d\omega$$

$$= \frac{A}{2j} \int_{-\infty}^{\infty} [\delta(\omega - \Omega) - \delta(\omega + \Omega)]\, |H(\omega)|\, e^{j\phi(\omega)}\, e^{j\omega t}\, d\omega$$

Due to the definition of the unit-impulse function, the integrand is zero except for frequencies $\omega = \pm \Omega$. Therefore:

$$y(t) = \frac{A}{2j} \int_{-\infty}^{\infty} \left[\delta(\omega - \Omega)\, M(\Omega)\, e^{j\phi(\Omega)} - \delta(\omega + \Omega)\, M(-\Omega)\, e^{j\phi(-\Omega)}\right] e^{j\omega t}\, d\omega$$

Making use of the symmetry of the response function, the output becomes:

$$y(t) = A M(\Omega) \sin [\Omega t + \phi(\Omega)] \qquad [1.24]$$

This equation shows that the response to a sinusoidal input is again a sinusoidal signal with a constant frequency, amplitude $A|H(\omega)|$ and phase $\phi(\omega)$. Equation [1.24] states what may be intuitively expected, that a linear system affects the amplitude and phase of a sinusoidal input signal, both effects being frequency-dependent. The frequency response function, $H(\omega)$, of a linear time-invariant system can be easily determined experimentally. We excite the input by a sinusoidal signal of angular frequency Ω and measure the corresponding output. The ratio of output and input amplitudes $A|H(\Omega)|/A$ gives then the amplitude response $|H(\omega)|$ for $\omega = \Omega$ and the difference between the output and input phase determines the phase response, $\phi(\omega)$, for the same frequency. Letting Ω vary over the frequency range of interest, amplitude and phase characteristics are measured as functions of ω. True steady-state signals are of course not realizable but in all practical applications, signal durations long enough to obtain reasonable results can be used.

The relation described by [1.24] may easily be extended to any input signal which can be resolved into a sequence of sinusoidal components, i.e. to signals for which the Fourier transform exists. We modify each of these components according to [1.24] and add them together. Due to the additive property of linear systems the sum of modified components yields the total output $y(t)$.

It follows immediately from [1.14], [1.15] and [1.24] that when a signal passes through a filter its form is influenced by the amplitude and phase responses of the filter. The functions $M(\omega)$ and $\phi(\omega)$ introduce frequency-dependent attenuation and phase shift, respectively, in the frequency components of the input signal. As a result, the form of the output signal will differ from that of the input signal. Phase characteristics evidently do not assist in suppressing the undesired frequency components and therefore their influence should be minimized.

As a direct consequence of the phase response there will be generally a time shift between the input and output signals. Furthermore, wavelets of the input containing different frequency components will be shifted in time with respect to each other at the output of the system. In other words, a frequency-dependent time shift has to be expected in the output of the filter. Naturally, this is a serious obstacle in most seismological applications. When analyzing records with weak seismic signals we are, of course, interested in an increase in the signal/noise ratio but not at a price of changing the arrival times of studied signals.

The problem now before us, is to determine conditions for *distortionless transmission*. A filter is considered to be distortionless if its response to any input has the same form as the input. Firstly, let us assume the simplest case when the output is a replica of the input except for a constant gain factor M. Then:

$$y(t) = Mx(t) \qquad [1.25]$$

In view of [1.15], this will be true only when $|H(\omega)| = M =$ constant and $\phi(\omega) = 0$, which means that the gain must be constant, and independent of frequency, and the phase must be equal to zero in the entire frequency band. Assuming [1.25], then, the frequency response function is real and constant for all frequencies. Therefore, the corresponding unit-impulse response, which is assumed to be a real function of time, must necessarily be an even function. However, since causality and evenness of the unit-impulse response, i.e. causality and *zero-phase response* are contradictory requirements, there is no causal system which could perfectly satisfy the assumption in [1.25].

Secondly, let us examine a case where the output is a delayed replica of the input:

$$y(t) = Mx(t - t_1) \qquad [1.26]$$

Note that the constant time delay, t_1, of the filter introduces a time shift between the input and output signals. Nevertheless, this shift must be the same for any frequency component, due to the unchanged form of the signal when passing through the filter. Consequently, this delay causes nothing more than a shift of the output time scale by the constant amount of t_1. The relative time delay may easily be estimated by cross-correlating the input and output signals, for example. The cross-correlation function will give a peak at time displacement equal to t_1.

Due to [1.14] we have:

$$Y(\omega) = X(\omega) H(\omega) = \int_{-\infty}^{\infty} Mx(t - t_1) e^{-j\omega t} dt$$

Introducing a change of variables $\tau = t - t_1$, we obtain:

$$Y(\omega) = M \int_{-\infty}^{\infty} x(\tau) \exp[-j\omega(\tau + t_1)] d\tau = M \exp(-j\omega t_1) X(\omega)$$

so that the response function becomes:

$$H(\omega) = M \exp(-j\omega t_1)$$

This result also follows directly from the well-known time-shifting theorem. Generally, for distortionless transmission the gain must be constant, frequency-invariant, and the phase response must be a linear function of frequency. If $|H(\omega)|$ is not a constant the transmission is *amplitude distorted*, if $\phi(\omega)$ is not a linear function of frequency the transmission is *phase-distorted*. Smith (1958) makes use of a seismic pulse passing through a given system to demonstrate the amplitude and phase distortion. Phase-distortionless filtering realized by means of a magnetic delay-line filter has been described by Domenico (1965).

Introducing the condition for the distortionless transmission into [1.12] for the corresponding unit-impulse response we have:

$$h(t) = \frac{1}{2\pi} \int_{-\infty}^{\infty} H(\omega) e^{j\omega t} d\omega = \frac{1}{2\pi} \int_{-\infty}^{\infty} M \exp[j\omega(t - t_1)] d\omega$$

Making use of Euler's formula and cancelling the imaginary part we obtain:

$$h(t) = \frac{1}{2\pi} \int_{-\infty}^{\infty} M \cos \omega(t - t_1) d\omega$$

As follows immediately from this expression, $h(t)$ is a symmetric function about $t = t_1$. In other words, the unit-impulse response of a *phase-distortionless filter* is a symmetric function of time. The response $h(t)$ is shifted in time by an amount of t_1 which is the slope of the linear phase characteristics. Since a constant shift applies to all frequencies it specifies also the time delay between the output and input of the filter.

1.5 ANALOG—DIGITAL CONVERSION

The voltage at the seismometer output, being a function of continuous time, has a purely analog character. The case for other geophysical quantities is similar. It is evident that geophysical data have to be converted into digital form prior to processing by digital computers. This is due to the character of any digital computer which for nonreal-time processing requires the input data in the form of a finite sequence of numbers. The process of converting an analog signal into a sequence of numbers is called *digitization*. It consists of two separate operations: *sampling* and *quantization*. Sampling defines the time instants at which the signal is to be observed, while quantization is the conversion of continuous signal amplitudes, at the sampling points, into a sequence of numbers. Throughout the present book, if not mentioned otherwise, the sampling is carried out at equally spaced intervals separated by the *sampling period* Δt. Nevertheless, in various geophysical applications, depending upon the character of observational data, digitization is first made in nonequally spaced instants and then, as the second step, interpolated at equally spaced points (see e.g. Naidu, 1968). As far as the quantization concerns, we shall here assume the quantizing error to be small enough to be neglected. Effects of errors generated during the conversion of analog geophysical records are described e.g. by An (1965) and Manzoni (1967).

The simplest digitizing consists in visual readings of enlarged records, but today various automatic and semiautomatic converters (see e.g. Belotelov and Rykunov, 1963; Wickens and Kollar, 1967) as well as digital seismographs (see e.g. De Bremaecker et al., 1962, 1963; Haubrich and Iyer, 1962; Miller, 1963; Phinney and Smith, 1963; Smith, 1965; Burke et al., 1970; Allsopp et al., 1972) are available and commonly utilized. Principles for converting an analog electrical signal (voltage) into its digital form have been discussed e.g. by Dobrin and Ward (1962).

Theoretically, digitization may be performed with sufficiently dense sampling which will guarantee good resemblance between the digital and related analog signal. However, sampling at closely spaced points yields correlated and highly redundant data and thus unnecessarily increases the total amount of data, labor and cost of computation. On the other hand, sampling at instants which are far apart may lead to confusion between the low- and high-frequency components in the original data. Evidently, the length of Δt is ruled by a reasonable compromise between the two opposing requirements. From our experience it follows that a good rule is to choose Δt to be about one tenth to one fifth of the periods studied (see also Melton, 1967).

It is important to emphasize that the similarity between the analog and digital forms may disappear when the signals are transformed into the frequency domain. This is true even if the resemblance in the time domain has been almost perfect. The apparent disagreement in the frequency domain has to be thoroughly considered in any digital-filter design. We shall discuss this problem in more detail below.

Since the technical realization of the analog–digital converter does not pertain to our discussion, we shall limit ourselves to an ideal device, to a *sampler* (Fig. 1.2a) with the following properties:

(1) The sampler can open and close instantly.
(2) The sampler operation is periodic, with sampling period Δt.
(3) The pulse duration, when the switch is closed, is small compared to the sampling period.

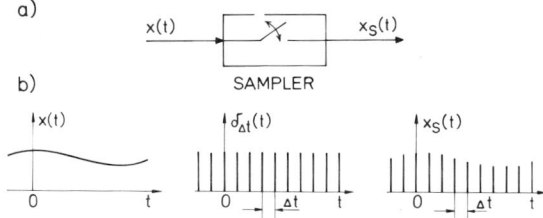

Fig. 1.2. (a) The sampling system; (b) the analog input, $x(t)$, ideal sampling function $\delta_{\Delta t}(t)$ and ideal-sampler output, $x_S(t)$.

It is evident that the sampler performs only the first operation of digitization, the sampling. Quantization is carried out by an additional digital measuring device (so-called *quantizer*, not shown in Fig. 1.2a), e.g. by a digital voltmeter. Let $x(t)$ be the analog input signal and $x_S(t)$ the sampler output. Further, let $\delta_{\Delta t}(t)$ be the ideal sampling function representing an infinite train of equidistant unit impulses. The sampling function then becomes:

$$\delta_{\Delta t}(t) = \sum_{n=-\infty}^{\infty} \delta(t - n\Delta t) \qquad [1.27]$$

where $\Delta t = $ constant is the time interval between two consecutive pulses. The output of the sampler is:

$$x_S(t) = x(t)\, \delta_{\Delta t}(t)$$

$$= x(t) \sum_{n=-\infty}^{\infty} \delta(t - n\Delta t) = \sum_{n=-\infty}^{\infty} x(n\Delta t)\, \delta(t - n\Delta t) \qquad [1.28]$$

where $x(n\Delta t) = \displaystyle\int_{-\infty}^{\infty} x(t)\, \delta(t - n\Delta t)\, dt$ is the value of the nth sample.

Equation [1.28] states that $x_S(t)$, which is a product of the ideal sampling function and the analog input, can be visualized as amplitude impulse modulation (Fig. 1.2b). It follows from the definition, [1.27], that $\delta_{\Delta t}(t)$ is a periodic function of time with period Δt. Hence, in terms of Fourier series $\delta_{\Delta t}(t)$ becomes:

$$\delta_{\Delta t}(t) = \sum_{n=-\infty}^{\infty} c_n \exp(jn\omega_s t) \qquad [1.29]$$

ANALOG–DIGITAL CONVERSION

where ω_s is the so-called *sampling angular frequency* in radians/sec:

$$\omega_s = 2\pi f_s = 2\pi/\Delta t$$

and c_n are Fourier coefficients defined as:

$$c_n = (1/\Delta t) \int_{-\Delta t/2}^{\Delta t/2} \delta_{\Delta t}(t) \exp(-jn\omega_s t) \, dt = 1/\Delta t$$

Here we note that within the interval $< -\Delta t/2, \Delta t/2 >$ the function $\delta_{\Delta t}(t)$ is nonzero only for $t = 0$, and that the pulse has a unit area. Substitution of c_n in [1.29] yields the ideal sampling function as:

$$\delta_{\Delta t}(t) = (1/\Delta t) \sum_{n-\infty}^{\infty} \exp(jn\omega_s t)$$

The input–output relation of an ideal sampler, given by [1.28], can now be rewritten in a form:

$$x_S(t) = (1/\Delta t) \sum_{n=-\infty}^{\infty} x(t) \exp(jn\omega_s t) \qquad [1.30]$$

Applying the frequency-shifting theorem we have a Fourier transform pair:

$$x(t) \exp(j\omega_s t) \longleftrightarrow X(\omega - \omega_s)$$

and finally, the transform of [1.30] gives:

$$X_S(\omega) = (1/\Delta t) \sum_{n=-\infty}^{\infty} X(\omega - n\omega_s) \qquad [1.31]$$

where $X(\omega)$ is the Fourier transform of the sampler input:

$$X(\omega) = \int_{-\infty}^{\infty} x(t) e^{-j\omega t} \, dt$$

Equation [1.31] demonstrates the fundamental consequences of sampling. While the frequency spectrum of the analog input of the sampler is a nonperiodic function, the spectrum of the sampler output is always a periodic function of frequency with a period equal to ω_s. That is:

$$X_S(\omega) = X_S(\omega + n\omega_s) \qquad [1.32]$$

where n is an integer. Generally speaking, discrete time functions, resulting from periodic sampling, possess periodic frequency spectra, while periodic time functions have discrete frequency spectra. Supposing that the spectrum of the signal input to the sampler is given, the spectrum of the output can readily be determined from [1.32]. Figure 1.3b shows an example of the output spectrum. It consists of *primary* and *complementary components*. The former are similar to those of the input spectrum (Fig. 1.3a) except for the normalizing factor of $1/\Delta t$; the latter are identical with those of the primary components, but they are displayed by $\pm n\omega_s$ units.

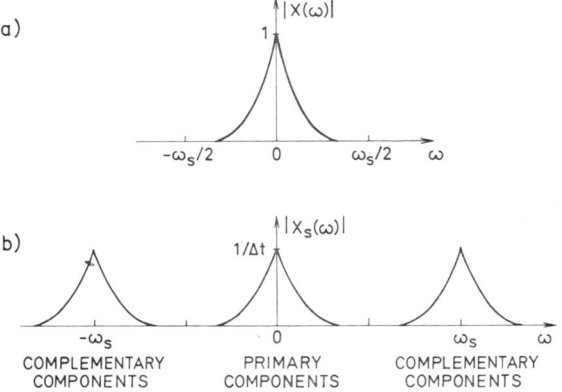

Fig. 1.3. Amplitude spectra: (a) of the analog input of the sampler; (b) of the sampled output. Δt is the sampling period.

Let the input to the sampler be band-limited, i.e. its spectrum is nonzero only over a finite range of frequency, say from $-\omega_c$ to ω_c, i.e. $|X(\omega)| = 0$ for $|\omega| > \omega_c$. We will show below that if this is the case then it is possible to recover the input signal from the spectrum of the sampler output, provided that $\omega_s/2 > \omega_c$. In other words the digitized signal is then equivalent to its analog form, and by digitizing we do not lose any information carried by the signal. The quantity $\omega_s/2 = \pi/\Delta t$ in radians/sec is called the *folding angular frequency* or *Nyquist angular frequency*.

The influence of the length of Δt on the behaviour of the spectrum $X_S(\omega)$ may be demonstrated by using a simple example shown in Fig. 1.4. Consider a continuous real and even time function, $x(t)$, which has a band-limited spectrum $X(\omega)$ diagrammed in Fig. 1.4a. The spectrum $X(\omega)$ is nonzero only within the frequency range $-\omega_c$, ω_c. The evenness of $x(t)$ has been introduced in order to make the spectrum $X(\omega)$ real which simplifies the graphical representation. Let the function $x(t)$ be digitized with a sampling frequency $\omega_s = 2\pi/\Delta t$ such that $\omega_s/2 > \omega_c$ (Fig. 1.4b). Then, apart from the normalizing factor $1/\Delta t$, the primary and complementary components are separate perfect copies of the input spectrum. Therefore, the full recovery of the input signal in terms of the sampler output spectrum is possible. When increasing the sampling period Δt up to the limit length when $\omega_s/2 = \omega_c$ (Fig. 1.4c) the picture will be similar. Any further increase of Δt will result in overlapping, so-called *aliasing*, of primary and complementary components (Fig. 1.4d). The recovery of the input signal from the distorted output spectrum is then no longer possible. Summarizing, we can state that a function whose spectrum is band-

ANALOG–DIGITAL CONVERSION

limited is fully specified by samples spaced at equal intervals not exceeding π/ω_c. In other words, the highest frequency which can be reproduced from data sampled with the sampling period Δt is:

$$f_N = f_s/2 = \omega_s/4\pi = 1/2\Delta t \qquad [1.33]$$

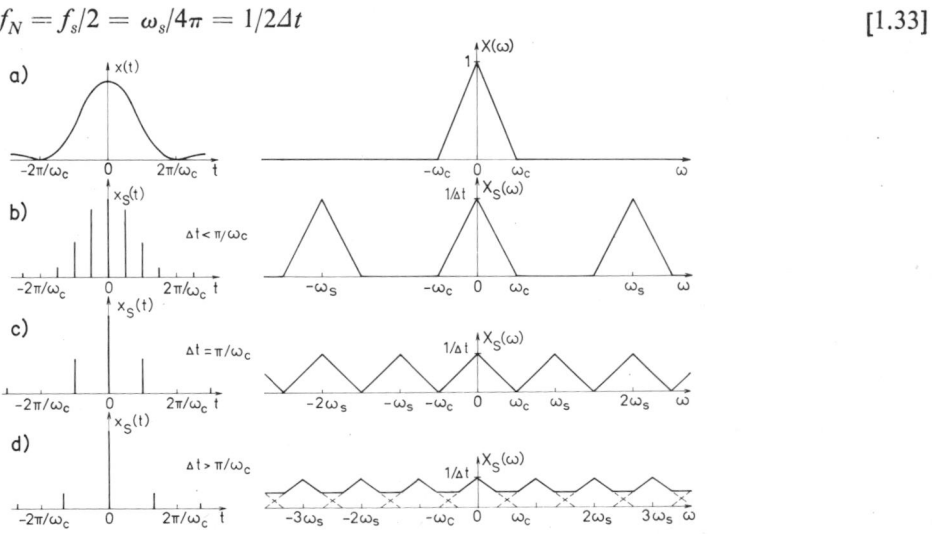

Fig. 1.4. Influence of the length of the sampling period upon the form of the spectrum at the sampler output (see text).

In agreement with the quantity $\omega_s/2$, f_N in cps is called the *folding frequency* or *Nyquist frequency*. Equation [1.33] represents an important result, commonly known as the *sampling theorem*. It says that at least two samples per cycle are required to define a frequency component in the original signal.

For most recorded geophysical signals the requirement of a band-limited spectrum is rather severe. Consequently, while processing seismic and similar signals we frequently meet the aliasing problem. Several methods are available to minimize its influence. The simplest is to choose Δt sufficiently small so that processed signals will have little energy above the associated folding frequency f_N. In general, reasonably good results may be received for f_N being at least twice as great as the highest predicted (anticipated) frequencies. However, small Δt increases the amount of data and the cost of computation. Another possibility is to prefilter the analog signal by a so-called *anti-aliasing filter* (Donnell, 1967). An ideal low-pass filter (see Chapter 3) cascaded with a sampler will cut off, prior to sampling, all components with frequencies above a prescribed limit. Here we suppose that the suppressed components are carrying unimportant information. The prefiltered signals may further be treated as band-limited. An alternative method to suppress aliasing effects is the usage of varying sampling periods (see also Båth, 1974, p. 151). In contrast to the properties of the sampler mentioned above this approach requires an aperiodic sampling and therefore will not be considered further.

1.6 z-TRANSFORM

The output of a sampler may be treated as though it were a sequence of numbers $x_n = x(n\Delta t)$, where n is an integer. Numbers x_n are generated by sampling the continuous input at instants separated by the sampling period Δt. This approach is convenient when the digital-computer processing is assumed. The Fourier transform of both sides of [1.28] yields:

$$X_S(\omega) = \sum_{n=-\infty}^{\infty} x_n\, e^{-j\omega n \Delta t}$$

and substitution of $z = e^{j\omega \Delta t}$ gives:

$$X(z) = \{x_n\} = \sum_{n=-\infty}^{\infty} x_n\, z^{-n} \qquad [1.34]$$

While $X_S(\omega)$ is a function of the real angular frequency, $X(z)$ is a function of the complex variable $z = e^{j\omega \Delta t} = \cos\omega\Delta t + j\sin\omega\Delta t$. The absolute value $|z| = 1$ for any angular frequency ω. Equation [1.34] defines the so-called *z-transform* under the restriction that the independent variable, z, lies on the unit circle. As ω passes from $-\infty$ to ∞, the complex variable z rotates along the unit circle in the z-plane. Any change of ω by $\pm 2\pi/\Delta t$ produces periodic duplication in the z-plane, which has to be viewed as a direct consequence of the process of digitization.

Consider, for example, that ω increases in steps by $\pi/2\Delta t$ from 0 to $2\pi/\Delta t$. Corresponding values of z are easily obtained from the relation $z = e^{j\omega \Delta t}$:

ω	0	$\pi/2\Delta t$	$\pi/\Delta t$	$3\pi/2\Delta t$	$2\pi/\Delta t$
z	1	j	-1	$-j$	1

Further increase of ω from ω_s to $2\omega_s$ will duplicate the given values of z and the process will be repeated for any increase or decrease of the angular frequency by a value of ω_s.

It follows immediately from the definition [1.34] that the z-transform of a sequence of time samples:

$$x = (\ldots, x_{-2}, x_{-1}, x_0, x_1, x_2, \ldots)$$

is a polynomial in z, with coefficients equal to the sampled signal values. Thus, in accordance with [1.34] we have:

$$X(z) = \ldots + x_{-2}z^2 + x_{-1}z + x_0 + x_1 z^{-1} + x_2 z^{-2} + \ldots =$$

$$= \sum_{n=-\infty}^{\infty} x_n\, z^{-n} \qquad [1.35]$$

The multiplication of the transform by a factor z^{-1} means physically nothing more than a shift of the sequence by one sampling period in the direction of increasing time. At times, the polynomial in [1.35] may be expressed in a finite form and for some of the commonly used time functions the transform may be derived analytically. Below, we present several examples. Lower case letters will be used to denote time sequencies, while capital letters will denote corresponding z-transforms.

Delta function. Consider the digital equivalent of the unit-impulse function:

$$x = (\ldots, 0, 0, \underset{\uparrow}{1}, 0, 0, \ldots)$$

where the vertical arrow indicates the time origin $t = 0$ (after Robinson, 1967b, p. 103). Then:

$$X(z) = \sum_{n=-\infty}^{\infty} x_n z^{-n} = 1$$

There is a clear analogy between the sequence $x = (1, 0, 0, \ldots)$ and the unit impulse $\delta(t)$ as they are utilized in the theory of digital and analog systems, respectively. Suppose now, that the impulse is delayed by, say, two sampling periods:

$$x = (\ldots, 0, \underset{\uparrow}{0}, 0, 1, 0, \ldots)$$

The transform becomes $X(z) = z^{-2}$.

Unit-step function.

$$x = (\ldots, 0, 0, \underset{\uparrow}{1}, 1, 1, \ldots)$$

$$X_1(z) = 1 + z^{-1} + z^{-2} + \ldots = (1 - z^{-1})^{-1} \quad \text{for } |z^{-1}| < 1$$

Consider the same time sequence advanced in time by two sampling periods:

$$x = (\ldots, 0, 0, 1, \underset{\uparrow}{1}, 1, 1, \ldots)$$

Then:

$$X_2(z) = z^2 + z + 1 + z^{-1} + z^{-2} + \ldots = z^2 X_1(z) \quad \text{for } |z^{-1}| < 1$$

Exponential function. Assuming sampling with period Δt, the sequence coefficients then are:

$$x_n = A e^{-\alpha n \Delta t} \quad n \geq 0$$

$$X(z) = A[1 + e^{-\alpha \Delta t} z^{-1} + e^{-2\alpha \Delta t} z^{-2} + \ldots] = A/(1 - e^{-\alpha \Delta t} z^{-1})$$
$$\text{for } |e^{-\alpha \Delta t} z^{-1}| < 1$$

Table II provides a short review of Laplace transforms, z-transforms and their inverse for most common time functions. For more extensive z-transform tables the reader is referred to Jury (1958).

It is worth mentioning that in addition to [1.34] which is considered to be the

TABLE II

Laplace transforms, z-transforms and their inverse of common time functions and several basic theorems

Description of the function	Time function	Laplace transform	z-transform	Time sequence
Unit impulse	$\delta(t)$	1	1	$\delta(n\Delta t) = \begin{cases} 1 \text{ for } n=0 \\ 0 \text{ for } n \neq 0 \end{cases}$
Unit step	$u(t) = \begin{cases} 1 \text{ for } t \geq 0 \\ 0 \text{ for } t < 0 \end{cases}$	$\dfrac{1}{p}$	$\dfrac{1}{1-z^{-1}}$	$u(n\Delta t) = \begin{cases} 1 \text{ for } n \geq 0 \\ 0 \text{ for } n < 0 \end{cases}$
Ramp	t	$\dfrac{1}{p^2}$	$\dfrac{\Delta t\, z^{-1}}{(1-z^{-1})^2}$	$n\Delta t$
Unit acceleration, parabola	$\tfrac{1}{2}t^2$	$\dfrac{1}{p^3}$	$\dfrac{\Delta t^2\, z^{-1}(1+z^{-1})}{(1-z^{-1})^3}$	$\tfrac{1}{2}(n\Delta t)^2$
Exponential decay	$e^{-\alpha t}$	$\dfrac{1}{p+\alpha}$	$\dfrac{1}{1 - e^{-\alpha \Delta t}\, z^{-1}}$	$e^{-\alpha n \Delta t}$
	$t\, e^{-\alpha t}$	$\dfrac{1}{(p+\alpha)^2}$	$\dfrac{e^{-\alpha \Delta t}\, z^{-1}}{(1 - e^{-\alpha \Delta t}\, z^{-1})^2}$	$n\Delta t\, e^{-\alpha n \Delta t}$

(continued)

TABLE II (continued)

Laplace transforms, z-transforms and their inverse of common time functions and several basic theorems

Description of the function	Time function	Laplace transform	z-transform	Time sequence
	$t^2 e^{-\alpha t}$	$\dfrac{2}{(p+\alpha)^3}$	$\dfrac{e^{-\alpha \Delta t} z^{-1}(1+e^{-\alpha \Delta t} z^{-1})}{(1-e^{-\alpha \Delta t} z^{-1})^3}$	$(n\Delta t)^2 e^{-\alpha n \Delta t}$
Unit step minus exponential decay	$1 - e^{-\alpha t}$	$\dfrac{\alpha}{p(p+\alpha)}$	$\dfrac{(1-e^{-\alpha \Delta t}) z^{-1}}{(1-z^{-1})(1-e^{-\alpha \Delta t} z^{-1})}$	$1 - e^{-\alpha n \Delta t}$
Sine wave	$\sin \Omega t$	$\dfrac{\Omega}{p^2 + \Omega^2}$	$\dfrac{z^{-1} \sin \Omega \Delta t}{1 - z^{-1} 2 \cos \Omega \Delta t + z^{-2}}$	$\sin \Omega n \Delta t$
Cosine wave	$\cos \Omega t$	$\dfrac{p}{p^2 + \Omega^2}$	$\dfrac{1 - z^{-1} \cos \Omega \Delta t}{1 - z^{-1} 2 \cos \Omega \Delta t + z^{-2}}$	$\cos \Omega n \Delta t$
Damped sine wave	$e^{-\alpha t} \sin \Omega t$	$\dfrac{\Omega}{(p+\alpha)^2 + \Omega^2}$	$\dfrac{z^{-1} e^{-\alpha \Delta t} \sin \Omega \Delta t}{1 - z^{-1} 2 e^{-\alpha \Delta t} \cos \Omega \Delta t + z^{-2} e^{-2\alpha \Delta t}}$	$e^{-\alpha n \Delta t} \sin \Omega n \Delta$
Damped cosine wave	$e^{-at} \cos \Omega t$	$\dfrac{p+a}{(p+a)^2 + \Omega^2}$	$\dfrac{1 - z^{-1} e^{-\alpha \Delta t} \cos \Omega \Delta t}{1 - z^{-1} 2 e^{-\alpha \Delta t} \cos \Omega \Delta t + z^{-2} e^{-2\alpha \Delta t}}$	$e^{-\alpha n \Delta t} \cos \Omega n \Delta t$
Linearity theorem	$x_1(t) \pm x_2(t)$	$X_1(p) \pm X_2(p)$	$X_1(z) \pm X_2(z)$	$x_1(n\Delta t) \pm x_2(n\Delta t)$
Translation theorem	$x(t - k\Delta t)$	$e^{-k\Delta pt} X(p)$	$z^{-k} X(z)$	$x(n\Delta t - k\Delta t)$
Scale-change theorem	$e^{\pm \alpha t} x(t)$	$X(p \mp a)$	$X(e^{\mp \alpha \Delta t} z)$	$e^{\pm n\Delta t} x(n\Delta t)$

engineering definition, the transform of the sequence $\{x_n\}$ is sometimes defined (so-called *Laplace definition*) as:

$$\tilde{X}(z) = \sum_{n=-\infty}^{\infty} x_n z^n$$

If not mentioned otherwise, the former definition will be used throughout the present book. The change from one to the other version is carried out simply by substituting z for z^{-1} but the two possibilities must be considered when referring to other books.

In many cases the sequence of time pulses is defined only for nonnegative values of n (see the example of an exponential function above), then the transform is called *one-sided transform*. Otherwise, as for example in [1.34], it is the *two-sided transform*.

The z-transform is not limited only to the unit circle, as defined in [1.34], but may be extended to the entire complex plane. In doing this, consider now an analog causal signal $x(t)$. Its sampled form, according to [1.28], is:

$$x_S(t) = \sum_{n=0}^{\infty} x_n \delta(t - n\Delta t) \qquad [1.36]$$

Instead of utilizing the Fourier transform as in the previous case, [1.34], let us now apply the Laplace transform to both sides of [1.36]. Then:

$$X_S(p) = \sum_{n=0}^{\infty} x_n \mathcal{L}\{\delta(t - n\Delta t)\}$$

Since $\mathcal{L}\{\delta(t)\} = 1$ and applying the time-shifting theorem:

$$\mathcal{L}\{x(t - n\Delta t)\} = e^{-p\Delta t} \mathcal{L}\{x(t)\}$$

the function $X_S(p)$ becomes:

$$X_S(p) = \sum_{n=0}^{\infty} x_n e^{-pn\Delta t}$$

where $p = \sigma + j\omega$ is a complex variable. Introducing:

$$z = e^{p\Delta t} = e^{\sigma\Delta t} e^{j\omega\Delta t} \qquad [1.37]$$

we have:

$$X(z) = \sum_{n=0}^{\infty} x_n z^{-n} \qquad [1.38]$$

Equation [1.38] describes a mathematical operation which is also commonly known as the *z-transform*. The transform is defined in the region of the complex plane where the series [1.38] converges. Provided that the series converges at least for one $z = z_0$, the sequence $\{x_n\}$ is z-transformable. Apart from the lower limits in the summations, [1.38] formally corresponds to [1.34]. The difference is that while in the former case the modulus $e^{\sigma \Delta t}$ may reach any value and the complex variable z is therefore defined in the entire complex plane, in the latter case the variable z is defined along the unit circle, $|z| = 1$, only. From this point of view the transform given by [1.34] may be considered as a special case of a more general z-transform defined by [1.38].

As follows from [1.38], the z-transform of the time function $x(t)$ depends on the function values at the sampling instants only. Addition of any time function which is zero at the sampling instants and takes an arbitrary nonzero value between these instants yields the same z-transform. Therefore, there are infinitely many time functions with identical z-transform, that is, the z-transform does not take into account variations of the analog function between sampling instants.

The inverse z-transform may be expressed in terms of a contour integral as:

$$x_n = \frac{1}{2\pi j} \int_C X(z) \, z^{n-1} \, dz \qquad \text{for } n = 0, 1, 2, \ldots$$

where the integration path, C, encloses all singularities of $X(z)$ and the direction of integration is counterclockwise. The path C has to be chosen so that it lies completely within the region of convergence of the series [1.38] and encloses the origin $z = 0$. The value of the contour integral is evaluated by the singularities of the integrand (see Båth, 1968). Thus, applying the *Cauchy theorem* we have:

$$x_n = \Sigma \text{ residues of } [X(z) \, z^{n-1}] \qquad \text{for } n = 0, 1, 2, \ldots$$

where the summation concerns all singularities encircled by C.

The design of effective digital filters frequently requires the use of noncausal time series $x(n\Delta t)$ nonzero for positive as well as negative n. Note, that in [1.38] we have considered a causal time series. Extending the z-transform also to noncausal functions gives a two-sided z-transform which is formally identical to [1.34]:

$$\mathcal{Z}\{x_n\} = X(z) = \sum_{n=-\infty}^{\infty} x_n \, z^{-n}$$

where it is assumed that the infinite series converges in the ring of convergence $R_1 < |z| < R_2$. The inverse of $X(z)$ is obtained by the same procedure as in the case of the one-sided z-transform. Provided that the convergence requirements are fulfilled, the integration can be performed along the unit circle, Γ. As in the preceding case we write:

$$x_n = \mathcal{Z}^{-1}\{X(z)\} = \frac{1}{2\pi j} \int_\Gamma X(z) \, z^{n-1} \, dz \qquad \text{for } -\infty < n < \infty$$

Since singularities of $X(z)$ may now lie inside as well as outside the integration path, the Cauchy theorem provides two possibilities in evaluating x_n, namely:

$x_n = \Sigma$ residue of $[X(z) z^{n-1}]$
all poles inside unit circle

and:

$x_n = -\Sigma$ residue of $[X(z) z^{n-1}]$
all poles outside unit circle

where both versions are valid for all n.

For more detailed treatment the reader may be referred to Jury (1964). Other methods for evaluating the sequence $\{x_n\}$ from the given $X(z)$ are discussed in later sections of the present book.

The substitution $z = e^{p\Delta t}$ may be considered as a mapping function which maps regions of the p-plane into regions of the z-plane. For the points on the imaginary axis of the p-plane, $\sigma = 0$ and the absolute value of z is unity. This implies, as we have already discussed, that the section of the imaginary axis of the p-plane lying between $\omega = 0$ and $\omega = 2\pi/\Delta t = \omega_s$ is mapped into the unit circle of the z-plane. Other parts of the imaginary axis just overlap on the unit circle. It follows from [1.37] that any line, $\sigma =$ constant, in the p-plane is mapped into the z-plane as a circle with its centre at $z = 0$ and radius $e^{\sigma \Delta t}$. For any point in the left halfplane of $p(\sigma < 0)$ 'e $^{\sigma \Delta t} < 1$ and $|z| < 1$. Hence, the primary strip of the p-plane $-\infty < \sigma < 0$, $-\omega_s/2 < \omega < \omega_s/2$, is mapped into the area inside the unit circle of the z-plane producing one sheet of the *Riemann surface* (see Båth, 1968). Complementary strips $-\infty < \sigma < 0$, $3\omega_s/2 < \omega < \omega_s/2$, etc. of the p-plane yield other sheets in the z-plane, with a common cut line on the negative real axis extending from $z = -1$ to $z = 0$ (Fig. 1.5). The right half of the p-plane ($\sigma > 0$, $|z| > 1$) is mapped into the area outside the unit circle in the z-plane. The engineering z-transform is related to that

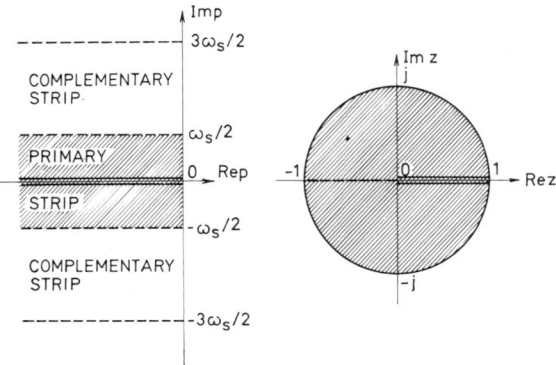

Fig. 1.5. Mapping from p-plane to z-plane.

defined by the Laplace definition by $X(z) = \tilde{X}(z^{-1})$. It is not difficult to check that the region inside the unit circle of the z-plane corresponds to the region outside the unit circle of the z^{-1}-plane.

It follows from the definition [1.34] that the transform limited to the unit circle in the z-plane is the Fourier transform of the sequence $\{x_n\}$. Assuming $\sigma = 0$, i.e. $|z| = 1$, and substituting $z = e^{j\omega \Delta t}$ back into [1.38] or directly from [1.34] we obtain the complex spectrum as a function of the real variable ω:

$$X(z) = X(e^{j\omega \Delta t}) = x_0 + x_1 e^{-j\omega \Delta t} + x_2 e^{-2j\omega \Delta t} + \ldots \quad [1.39]$$

This important phenomenon may be utilized e.g. in a graphical determination of $X(e^{j\omega \Delta t})$. Due to [1.39] the complex spectrum may be determined, at a certain frequency, $\omega = \Omega$, as a sum of a vector series. We may try the procedure numerically by making use of a simple example. Given the sequence:

$$x = (\ldots, 0, 0, 0, \underset{\uparrow}{1}, 2, 3, 4, 0, 0, \ldots)$$

find the complex spectrum $X(z)$ for $\omega = 0, \pi/2\Delta t, \pi/\Delta t$ and $3\pi/2\Delta t$. Similar ramp pulse has been also treated by Huang (1966). According to [1.39] the spectrum for $\omega = 0$ is given by a simple scalar summation:

$$X(1) = 1 + 2 + 3 + 4 = 10$$

since all the exponential terms in [1.39] become real and equal to 1. For the higher frequencies we obtain:

$$\omega = \pi/2\Delta t, \quad X(e^{j\omega \Delta t}) = 1 + 2e^{-j\pi/2} - 3 + 4e^{-j3\pi/2} = -2 + 2j$$
$$\omega = \pi/\Delta t, \quad X(e^{j\omega \Delta t}) = 1 + 2e^{-j\pi} + 3 + 4e^{-j\pi} = -2$$
$$\omega = 3\pi/2\Delta t, \quad X(e^{j\omega \Delta t}) = 1 + 2e^{-j3\pi/2} - 3 + 4e^{-j\pi/2} = -2 - 2j$$

The amplitude and phase spectra for given frequencies ω are:

ω	amplitude	phase
0	10	0
$\pi/2\Delta t$	$2\sqrt{2}$	$3\pi/4$
$\pi/\Delta t$	$\sqrt{2}$	$-\pi$
$3\pi/2\Delta t$	$2\sqrt{2}$	$-3\pi/4$

Amplitude and phase spectra for the frequency $\omega = 2\pi/\Delta t$ will repeat the values already received for $\omega=0$, due to the periodicity of the spectrum (note that $2\pi/\Delta t = \omega_s$). If the Laplace definition were applied, the amplitude spectrum would be the same but the phase spectrum would be the negative of that obtained above.

1.7 INPUT–OUTPUT RELATIONSHIPS FOR DIGITAL SYSTEMS

As a *digital system* we shall consider one in which both input and output signals are given as a sequence of discrete values. As follows from [1.6] the convolution integral describes the input–output relationship for analog signals. Let us suppose that the analog signal $x(t)$ is first sampled with the sampling period Δt and then applied to the input of the system (Fig. 1.6a). The input values at sampling instants $n\Delta t$ are $x_n = x(n\Delta t)$ where n is an integer. Similarly to the procedure applied previously to analog systems we may consider the system responses of individual input samples

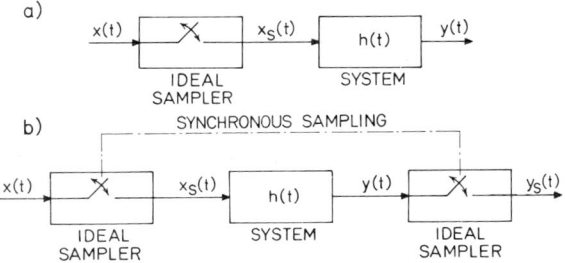

Fig. 1.6. Basic sampled-data systems: (a) sampled input and analog output; (b) both the input and output in sampled form.

separately. The total output is then the sum of all individual responses. It follows from the convolution integral that the response to the input sample x_n applied at the instant $t = n\Delta t$ is $\Delta t\, x_n\, h(t - n\Delta t)$ where $h(t)$ is the impulse response of the system. Individual samples of the input:

$$x = (\ldots, x_{-n}, x_{-n+1}, \ldots, x_{-2}, x_{-1}, x_0, x_1, x_2, \ldots, x_{n-1}, x_n, \ldots)$$

produce the following separate responses:

·
·
·

$\Delta t\, x_{-n}\, h(t + n\Delta t)$
$\Delta t\, x_{-n+1}\, h(t + n\Delta t - \Delta t)$

·
·
·

$\Delta t\, x_{-1}\, h(t + \Delta t)$
$\Delta t\, x_0\, h(t)$
$\Delta t\, x_1\, h(t - \Delta t)$

·
·
·

$\Delta t\, x_{n-1}\, h(t - n\Delta t + \Delta t)$
$\Delta t\, x_n\, h(t - n\Delta t)$

·
·

INPUT–OUTPUT RELATIONSHIPS FOR DIGITAL SYSTEMS

The total output becomes:

$$y(t) = \Delta t \sum_{n=-\infty}^{\infty} x_n h(t - n\Delta t)$$

Note, that while the input is of digital form the output of the system is a continuous function of time. Assuming further that the output is digitized synchronously with the input signal (see Fig. 1.6b), we obtain a reasonable approximation of a digital system. Thus, the output at the time instant $t = k\Delta t$ may be written as:

$$y_k = y(k\Delta t) = \Delta t \sum_{n=-\infty}^{\infty} x_n h(k\Delta t - n\Delta t) \qquad [1.40]$$

For the sake of simplicity from now on, if not mentioned otherwise, the sampling period Δt will be assumed to be unity. That means that the signal amplitude is strictly a function of index. The abbreviated form of [1.40] yields:

$$y_k = \sum_{n=-\infty}^{\infty} x_n h_{k-n} \qquad [1.41]$$

Evidently, due to the digital character of signals, the input–output relationship is described by a *convolution summation*. No special limitations have been required for sequences x and h in [1.41]. For a noncausal system, output values y_k will depend not only on the past but also on the future samples of the input sequence. When the input and/or the unit-impulse response is an infinitely long steady-state signal, then the output is also a sequence of infinite length.

Consider now another case, namely a one-sided input and a causal unit-impulse response both of infinite length. The two sequences are:

$x = (\ldots, 0, 0, x_0, x_1, x_2, \ldots)$
$h = (\ldots, 0, 0, h_0, h_1, h_2, \ldots)$

where the first nonzero samples correspond to the time instant $t = 0$. The output sequence is then given by:

$$y_k = \sum_{n=0}^{\infty} x_n h_{k-n} \qquad [1.42]$$

Due to the causal property and following the notation in [1.42], it is evident that the values y_k depend on present and past input values only. Causality also assures that the summation [1.42] will never contain more than $k + 1$ products of the type $x_n h_{k-n}$. Since the response of the filter is of infinite length, the output will also be a sequence of infinite length.

In geophysical applications the unit-impulse response usually has a form of decaying oscillations. Such responses may be *truncated*, i.e. they may be considered within a finite time interval only, without decreasing significantly the accuracy of the signal processing. The case is similar for a large category of input signals. For example, ground vibrations caused by distant seismic events of a reasonable magnitude produce a sudden increase of amplitudes on seismic records with well-defined onsets. These amplitudes remain relatively large within a certain time interval, then decay until they completely disappear in the background noise. Thus, records of earthquakes may be considered as transient waveforms with distinct onsets and vaguely defined terminations. On the other hand there are seismic signals of infinite length, such as microseismic noise, whose nearly stationary character makes it impossible to mark the beginning and the end of the signal. While signals from earthquakes and the impulse responses of causal, stable systems may be considered as one-sided and of finite length, the microseismic noise represents a typical two-sided, signal of infinite length.

The one-sided filter response truncated at the time instant $t = M\Delta t$:

$$h = (..., 0, 0, h_0, h_1, ..., h_{M-1}, h_M, 0, 0, ...)$$

and the input sequence:

$$x = (..., 0, 0, x_0, x_1, ..., x_{N-1}, x_N, 0, 0, ...)$$

will produce an output:

$$y_k = \sum_{n=0}^{N} x_n h_{k-n} \quad \text{for } 0 \leq k \leq M + N \qquad [1.43]$$

of finite length $M + N + 1$. The length of the output is one time unit less than the length of the input plus the filter response. The filtered signal is never shorter than the corresponding input.

Transforming both sides of [1.43] we find:

$$Y(z) = \sum_{k=0}^{M+N} y_k z^{-k} = \sum_{k=0}^{M+N} \sum_{n=0}^{N} x_n h_{k-n} z^{-k} \qquad [1.44]$$

Substitution $k - n = l$ reduces [1.44] into:

$$Y(z) = \sum_{l=0}^{M} \sum_{n=0}^{N} x_n h_l z^{-(l+n)}$$

Rearranging this yields:

$$Y(z) = \sum_{n=0}^{N} x_n z^{-n} \sum_{l=0}^{M} h_l z^{-l}$$

which according to the definition [1.38] is a product of z-transforms of the input and the system response sequences. Hence, the z-transform of the convolution summation or the z-transform of the output sequence is:

$$Y(z) = X(z) H(z) \qquad [1.45]$$

where $X(z)$ is the transformed input sequence and $H(z)$, called the *pulse-transfer function* or *system function*, is the z-transform of the unit-impulse response sequence. It is the same relation as for Fourier or Laplace transforms. The complex values of $H(z)$ when evaluated on the unit circle in the z-plane, i.e. for $|z| = |e^{j\omega \Delta t}| = 1$, provide the frequency response of the digital system. Equations [1.43] and [1.45] describe important input–output relationships for digital systems in the time and frequency domain, respectively. Convolution, [1.43], in the time domain corresponds to multiplication, [1.45], in the frequency domain. These conclusions show close analogy between analog and digital signals. The convolution of the two sequences x and h may be carried out in several different ways. Making use of simple wavelets we shall demonstrate the process of convolution in terms of a digital integration, two different graphical operations and an algebraic operation based on the z-transform.

Digital integration. Let the input signal be the sequence:

$x = (1, 2, 3, 4)$

and let the unit-impulse response of the filter be:

$h = (2, 1)$

The input and the impulse response are both one-sided finite sequences of lengths 4 and 2, respectively. The output sequence y will contain five $(4 + 2 - 1)$ terms. According to [1.43] individual members of the output sequence are:

$y_0 = x_0 h_0 = 2$
$y_1 = x_0 h_1 + x_1 h_0 = 1 + 4 = 5$
$y_2 = x_1 h_1 + x_2 h_0 = 2 + 6 = 8$
$y_3 = x_2 h_1 + x_3 h_0 = 3 + 8 = 11$
$y_4 = x_3 h_1 = 4$

Movable strip. The convolution of the sequences x and h may be easily performed by making use of a movable strip. First we reverse the sequence h and then we write x and reversed h on two strips in the following way:

$h_1 \quad h_0 \quad \rightarrow$
$\quad\quad x_0 \quad x_1 \quad x_2 \quad x_3$

Suppose the strip with h_0 and h_1 terms is movable with respect to the other strip. The output value y_0 is obtained by multiplying vertical pairs and summing. To obtain y_1 we move the h-strip one position towards the right and perform vertical multiplications. Summing these products we obtain the value of y_1. Repeating the operation for further shifts of the h-strip, we obtain the remaining output values. It is evident that for reasonably short signals the process of convolution may quickly be carried out with nothing more than a simple desk calculator.

In general, the movable-strip approach reminds us of the cross-correlation between x and h. Displacement, multiplication and summation (digital integration) are operations involved in evaluating both the convolution and cross-correlation. The essential difference is that convolution also involves folding, whereas cross-correlation does not. However, if h is a digital version of an even time function, the effect of the folding would disappear and the convolution and cross-correlation would yield the same results.

Folding. The input and impulse-response sequences together with corresponding products may be written in a form as shown below:

$$
\begin{array}{cccc}
x_0 & x_1 & x_2 & x_3 \\
h_0\ x_0 h_0 & x_1 h_0 & x_2 h_0 & x_3 h_0 \\
h_1\ x_0 h_1 & x_1 h_1 & x_2 h_1 & x_3 h_1
\end{array}
$$

Summing the products within diagonal strips we obtain:

1. strip $x_0 h_0$ $= 2$
2. strip $x_0 h_1 + x_1 h_0 = 5$
3. strip $x_1 h_1 + x_2 h_0 = 8$
4. strip $x_2 h_1 + x_3 h_0 = 11$
5. strip $x_3 h_1$ $= 4$

Comparison with [1.43] shows that diagonal-strip summations provide the values of the output sequence. Thus introducing numerical values, diagonal strips yield the sequence $y = (2, 5, 8, 11, 4)$. The whole process may be viewed as a successive folding of the diagonal strips (Robinson, 1967b, p. 114).

Multiplication of polynomials. In general, z-transforms of the input and filter impulse response are given by the power series:

$$X(z) = x_0 + x_1 z^{-1} + x_2 z^{-2} + \ldots$$
$$H(z) = h_0 + h_1 z^{-1} + h_2 z^{-2} + \ldots$$

Introducing these power series into [1.45] we write:

$$Y(z) = \sum_{k=0}^{\infty} y_k z^{-k} = (x_0 + x_1 z^{-1} + x_2 z^{-2} + \ldots)(h_0 + h_1 z^{-1} + h_2 z^{-2} + \ldots)$$
$$= x_0 h_0 + (x_0 h_1 + x_1 h_0) z^{-1} + (x_0 h_2 + x_1 h_1 + x_2 h_0) z^{-2} + \ldots \quad [1.46]$$

Comparison of coefficients with the same power of z on both sides of [1.46] gives:

$$y_0 = x_0 h_0$$
$$y_1 = x_0 h_1 + x_1 h_0$$

.
.
.

where $y_0, y_1, \ldots$ are the samples of the output sequence. Thus, the convolution may be carried out as a multiplication of power series. Multiplication of the two

STABILITY OF DIGITAL SYSTEMS

z-transforms is, as shown in [1.45], an operation performed in the frequency domain. The return into the time domain is accomplished via the comparison of coefficients from both sides of [1.46]. In our example, numerical values give:

$$Y(z) = (1 + 2z^{-1} + 3z^{-2} + 4z^{-3})(2 + z^{-1}) = 2 + 5z^{-1} + 8z^{-2} + 11z^{-3} + 4z^{-4}$$

thus:

$$y = (2, 5, 8, 11, 4)$$

1.8 STABILITY OF DIGITAL SYSTEMS

In this chapter we have already expressed the essential stability requirement, that a stable system must respond with a bounded output to any bounded input. The stability property is usually looked at in terms of the corresponding system function and/or unit-impulse response. Such an approach is more convenient and frequently used in the filter design. There are a number of techniques such as the *Routh test, Nyquist criterion*, etc. (see e.g. Del Toro and Parker, 1960) which have been developed for investigating the stability conditions of linear analog systems. After certain modification, some of the techniques originally used for analog systems may be applied as stability tests for digital systems as well. Here, we shall discuss in more detail a criterion based on the correspondence between analog and digital systems.

Consider a transfer function $H(p)$ in the form of a ratio of two polynomials in p. The decisive stability factor is the location of poles $p_i = \sigma_i + j\omega_i$ of $H(p)$ in the complex p-plane. In general, poles p_i occur on the real axis or in complex conjugate pairs and may be of any order. The essential condition for a system to be stable is rather simple: the system is stable if all the poles p_i lie in the left half of the p-plane with $\operatorname{Re} p_i = \sigma_i < 0$. If this is the case, the output signal will contain only components which decay exponentially with increasing time thus ensuring a bounded response to any bounded input.

An analogous approach may be applied when the stability conditions of a digital system are studied. Since a system function $H(z)$ is a function of the complex variable z, the stability requirements may be expressed in terms of the location of poles in the z-plane. The zeros do not influence the stability of the system and may lie anywhere. Applying the results following from the analysis of analog systems we locate the poles in the z-plane and search for the area which corresponds to the left half of the p-plane. Care should be taken with the z-transform definitions, especially when results from various books are to be compared. Further, the function $H(z)$ may be expressed as a function of the independent variable z as well as z^{-1}. Consequently, the stability of the system may be tested in both the z- and z^{-1}-planes. Any of the procedures mentioned must of course provide the same answer since we change only the method and not the system itself.

In the extensive literature (see e.g. Bracewell, 1965; Robinson, 1967b; Ackroyd, 1973; Kanasewich, 1973) usually two definitions of the z-transform (see Table III)

TABLE III

Comparison of the two z-transform definitions

Definition	Variable	Some references
$H(z) = \sum_{n=0}^{\infty} h_n z^{-n}$	$z = e^{p\Delta t}$	Bracewell (1965), Ackroyd (1973), present book
$H(z) = \sum_{n=0}^{\infty} h_n z^{n}$	$z = e^{-p\Delta t}$	Robinson (1967b), Kanasewich (1973)

appear. In Fig. 1.5 we displayed the mapping process between the p- and z-planes. Provided that $z = e^{p\Delta t}$, the primary and all complementary strips which build up the left half of the p-plane are mapped into the z-plane as an infinite series of surfaces of the unit-circle interior. Thus, the "stable poles" in the p-plane correspond to poles inside the unit circle in the z-plane. If the poles are:

$$z_i = \exp(p_i \Delta t) = \exp(\sigma_i \Delta t) \exp(j\omega_i \Delta t)$$

then the stability condition for a digital system may be expressed by the inequality:

$$|z_i| = \exp(\sigma_i \Delta t) < 1 \quad \text{for all } i$$

When the complex variable $z = e^{-p\Delta t}$ is used instead, we have the alternative stability condition, namely:

$$|z_i| = \exp(-\sigma_i \Delta t) > 1 \quad \text{for all } i$$

which means that all poles in the z-plane have to be located outside the unit circle to ensure the stability of the system. Two other possibilities may arise when the z^{-1}-plane is also considered. The stable regions for all four cases are compared in Fig. 1.7.

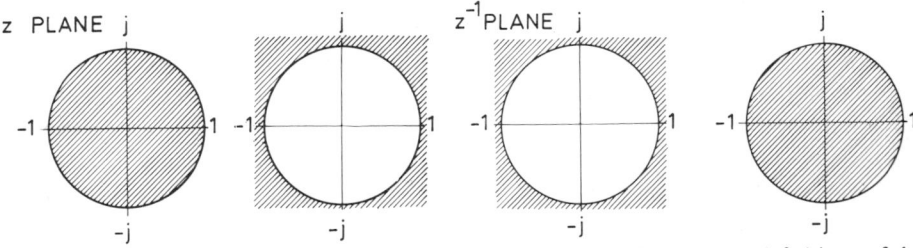

Fig. 1.7. Stability regions (shaded areas) in the z- and z^{-1}-planes for the two definitions of the variable z.

Chapter 2

DESIGN PRINCIPLES OF DIGITAL FILTERS

Digital operators transforming the input sequence into a corresponding output sequence are principally of two types, called nonrecursive and recursive, depending upon the method applied for the output-sequence determination.

Nonrecursive filtering or *transversal filtering*. The output depends solely on the applied input and the filter characteristics. The algorithm for computing the output sequence is given by the convolution summation in [1.43]. The system function, $H(z)$, of a nonrecursive filter, also called *convolution filter*, is a finite polynomial in z^{-1}, i.e. $H(z)$ has only zeros and no poles. Since $H(z)$ contains a finite number of terms, the corresponding unit-impulse response h is of finite duration. Nonrecursive filters are also called *finite-memory filters*.

Recursive filtering or *feedback filtering*. The output depends on the applied input, filter characteristics and the output of the filter. The system function of recursive filters is expressed as a ratio of two polynomials in z^{-1}. The long division generally provides a power series of infinite length. Consequently, recursive filters refer to infinite-duration response and are sometimes called *infinite-memory filters*. Usually, recursive filters are more efficient in that they demand less computing time when compared with convolution filters.

Besides recursive and nonrecursive filtering the present chapter outlines various design techniques for digital systems and presents discussion on quantization effects in digital filters.

2.1 NONRECURSIVE FILTERING

As follows from [1.45] the system function may generally be expressed as a ratio of two polynomials in z or z^{-1} so that:

$$H(z) = \frac{g_0 + g_1 z^{-1} + g_2 z^{-2} + \ldots + g_M z^{-M}}{f_0 + f_1 z^{-1} + f_2 z^{-2} + \ldots + f_L z^{-L}} = \frac{\sum_{k=0}^{M} g_k z^{-k}}{\sum_{k=0}^{L} f_k z^{-k}} \qquad [2.1]$$

where the values of coefficients g_k and f_k determine whether the filtering process will be of recursive or nonrecursive type. In general, the numerator and denominator in [2.1] may contain positive as well as negative exponents. If at least one of the g_k coefficients is nonzero and further if $f_0 \neq 0$ but all remaining f_k coefficients are zero, then the output sequence depends solely upon the present and past input terms and the filtering itself is of nonrecursive type. On the other hand, if at least one g_k coefficient is nonzero, $f_0 \neq 0$ and at least one more f_k coefficient is nonzero then the output sequence is expressed in terms of the present and past input and past output samples. The filtering then is of recursive or feedback type. Filter synthesis consists in choosing proper values of the g_k and f_k coefficients so as to achieve the desired behaviour of the filter. Recursive filters may be converted into corresponding nonrecursive filters simply via long division. The inverse conversion, however, requires rather laborious processing (see e.g. Shanks, 1967) and is more of theoretical importance.

In order to carry out the convolution summation in the nonrecursive filtering operation, the impulse-response sequence, h, must be available. Since the specification of filtering requirements directly in terms of the unit-impulse response is rather difficult, it is more common to derive $h(t)$ or h from the corresponding functions $H(p)$ or $H(z)$, respectively. In [2.1] the function $H(z)$ is written as a ratio of two polynomials and consequently the convolution described by [1.43] cannot be directly applied. One possible way to overcome this obstacle is to carry out the long division of [2.1], expanding $H(z)$ into a polynomial of z^{-1}:

$$H(z) = h_0 + h_1 z^{-1} + h_2 z^{-2} + \ldots$$

Generally, the power series $H(z)$ will be of infinite length, i.e. there will be an infinite number of nonzero coefficients $h_0, h_1, h_2, \ldots$, which is an inconvenient form for computer processing. The stability condition for such a digital system is easily obtained from [1.9] by substituting a summation for the integration. Thus for stable filters, the impulse response satisfies the inequality:

$$\sum_{n=-\infty}^{\infty} |h_n| < \infty \qquad [2.2]$$

and consequently the members of the sequence h converge to zero. If this is the case, the system function may be approximated to any prescribed accuracy by a polynomial in z^{-1} with a finite number of terms, such as:

$$H(z) \simeq \tilde{H}(z) = h_0 + h_1 z^{-1} + \ldots + h_N z^{-N}$$

Coefficients of the power series $\tilde{H}(z)$ approximate the impulse response h and therefore may be used in the convolution summation in [1.43] in order to obtain the filtered signal.

Consider a digital filter defined by its approximate system function $\tilde{H}(z)$ excited by an input sequence x. According to [1.43] the corresponding approximation of the output sequence is:

$$\tilde{y}_n = \sum_{k=0}^{N} x_k \tilde{h}_{n-k} \qquad [2.3]$$

where the symbol $\tilde{h} = (h_0, h_1, \ldots, h_N)$ is used for the truncated unit-impulse response. On one hand, a longer duration of $\tilde{h}$ gives a better approximation of $H(z)$ and of y_n (see e.g. Jones et al., 1955). On the other hand, the increasing length of $\tilde{h}$ will decrease the efficiency of the convolution operation indicated in [2.3] due to the increasing computing time required for the determination of any particular output term y_n. Evidently, efficiency and accuracy of the output determination are opposing requirements here and one has to choose a reasonable compromise. Below, we outline several design possibilities for finite-memory filters.

In Section 1.2.2 it has been shown that for analog systems the frequency response function, $H(\omega)$, and the corresponding unit-impulse response, $h(t)$, form a Fourier transform pair, provided that infinite integration limits are applied (see e.g. [1.11]). Consider time now as a discrete instead of continuous independent variable, i.e. a noncausal $h(t)$ is defined only at time instants $t = n\Delta t$ where n ranges over all integers from $-\infty$ to ∞. The Fourier transform of the infinitely long sequence h yields an approximation of $H(\omega)$ in the frequency range $-\omega_s/2 \leq \omega \leq \omega_s/2$ (see also Fig. 1.4). Generally, the approximation $\tilde{H}(\omega)$ will resemble $H(\omega)$ better for frequencies more distant from the folding point. In other words, the error $e(\omega) = \tilde{H}(\omega) - H(\omega)$ will be a function of frequency. Since infinitely long sequencies cannot be processed by computers let us consider a truncated impulse response of finite length $2N + 1$, i.e. $-N \leq n \leq N$. Assuming that $h(t)$ and Δt remain unchanged, the error of approximation is also influenced by the total number of terms in h, so that $e(\omega, N) = \tilde{H}(\omega) - H(\omega)$. Ormsby (1961) studied the behaviour of $e(\omega, N)$ for a special class of filters (see also Chapter 3). His empirical results make it possible to determine the minimum N necessary for a desired accuracy of approximation. For this class of digital filters $N \simeq 80$ usually provides an approximation with $e(\omega, N) \leq 1.5\%$. Because large errors occur near the discontinuities in $dH(\omega)/d\omega$ another possibility to decrease $e(\omega, N)$ is to eliminate these discontinuities. In doing this, Ormsby introduced parabolic smoothing of the original $H(\omega)$ in the neighbourhood of the slope discontinuities. Martin-Graham filters, discussed in the following chapter, may also provide useful results. Time-domain approach applied to barometric data has been employed by Galli and Randi (1967).

Another interesting method of designing a nonrecursive filter is the Fourier-series approach outlined by Kaiser (1966) among others. Let the transfer function, $H(\omega) = \mathcal{F}\{h(t)\}$, of an analog filter be defined in the frequency range $-\omega_s/2 \leq \omega \leq \omega_s/2$. Assuming further a digitized version of $h(t)$ and taking into account both the primary and complementary components, $H(\omega)$ becomes a periodic function of frequency with a period of $\omega_s = 2\pi/\Delta t$ and amplitudes reduced by a factor of $1/\Delta t$.

The amplitude response, $|H(\omega)|$, provides the best information about the frequency selectivity of the filter. The function $|H(\omega)|$ may be expanded into an infinite Fourier series over the frequency band considered (see also Aguilera et al., 1970) so that:

$$|H(\omega)| = \frac{a_0}{2} + \sum_{n=1}^{\infty} a_n \cos \omega n \Delta t$$

where the nth coefficient is given by:

$$a_n = \frac{2}{\omega_s} \int_{-\omega_s/2}^{\omega_s/2} |H(\omega)| \cos(\omega n \Delta t) \, d\omega$$

Making use of the substitution $z = e^{j\omega \Delta t}$ (the variable z moves along the unit circle in the z-plane) every cosine term of the series may be written as:

$$\cos n\omega \Delta t = (z^n + z^{-n})/2$$

The system function of the nonrecursive digital filter which approximates the original analog system $H(\omega)$ becomes:

$$H(z) = \frac{a_0}{2} + \frac{1}{2} \sum_{n=1}^{\infty} a_n (z^n + z^{-n}) = \frac{1}{2} \sum_{n=-\infty}^{\infty} a_n z^n \qquad [2.4]$$

where $a_{-n} = a_n$. Note, that the impulse-response values $h_n = a_n/2$, for any n.

Several interesting properties of the filter follow directly from [2.4]. Firstly, the filter no longer represents a causal system due to the nonzero values with negative time indexes in the impulse-response sequence. The impulse response consists of *anticipation* as well as of *memory terms* and consequently the output values are influenced by past, present and future input terms. This requires the input signal be available in a stored form. When the processing is carried out by computer in nonreal time this requirement usually does not create any difficulties. Secondly, the impulse-response terms form an even time sequence, i.e. anticipation and memory terms are mirror images of each other, and the filter is therefore a *symmetric* one. As a direct consequence of this symmetry, the imaginary part of $H(z)$ vanishes, $H(e^{j\omega \Delta t}) = |H(e^{j\omega \Delta t})|$, and the filter has zero-phase characteristics. Thus, a system described by [2.4] performs a phase-distortionless transmission (see also Section 1.4), a property usually required in geophysical applications. Kaiser (1966) also suggested the approximation which makes use of an infinite sine series. The choice between a cosine and sine series is determined by the behaviour of $H(\omega)$ at frequencies close to zero.

To illustrate the Fourier-series approach for nonrecursive filtering, consider a simple symmetric filter:

$$H(\omega) = \begin{cases} M & \text{for } -\omega_c \leq \omega \leq \omega_c, \quad |\omega_c| < |\omega_s/2| \\ 0 \end{cases}$$

The filter $H(\omega)$ is a low-pass filter, having constant gain, $|H(\omega)| = M$, in the low-frequency range $|\omega| \leq \omega_c$ and zero gain outside this range. The coefficients a_n are:

$$a_n = \frac{2A}{\omega_s} \int_{-\omega_c}^{\omega_c} \cos(\omega n \Delta t) \, d\omega = \frac{2M \omega_c \Delta t}{\pi} \frac{\sin \omega_c n \Delta t}{\omega_c n \Delta t}$$

As intuitively expected, coefficients a_n depend upon the shape of the filter and the sampling period. Substituting the values a_n back into [2.4] we obtain the system function of the desired filter. Since infinitely long signals cannot be processed by digital computers, the sequence h must be truncated. Unfortunately, the convergence of the sequence is usually not very rapid and rather large values of n are necessary in order to reach sufficiently small coefficients h_n. If we consider, for example, the sampling period $\Delta t = 0.1$ sec and the cut-off frequency $\omega_c = \pi$ rad/sec ($f_c = 0.5$ cps) then 55 terms ($n = 27$) of the impulse-response sequence are required to bring the values of coefficients h_n below 10% of the maximum value h_o.

Truncation introduces relatively large errors of approximation in the vicinity of discontinuities in $H(\omega)$ due to the *Gibbs phenomenon* (see e.g. Ormsby, 1961). Nevertheless the truncation effect may be minimized by modifying the terms h_n in a special way. Instead of h_n consider weighted values $w(n\Delta t)h_n$. Suppose the weighting function to be even, $w(t) = w(-t)$, and time limited, $w(t) = 0$ for $t > N\Delta t$. Then the weighted terms $w(n\Delta t)h_n$ form a finite sequence with nonzero values only for $-N \leq n \leq N$. Weighted coefficients yield the system function in the form:

$$H(z) = \frac{1}{2} h_0 w(0) + \frac{1}{2} \sum_{n=1}^{N} h_n w(n\Delta t) [z^{-n} + z^n] = \frac{1}{2} \sum_{n=-N}^{N} w(n\Delta t) h_n z^n$$

Multiplication by $w(t)$ in the time domain corresponds to convolution with $W(\omega)$ in the frequency domain, where $w(t)$ and $W(\omega)$ form a Fourier-transform pair. Provided that the convolution with $W(\omega)$ does not significantly influence the desired filtering properties of $H(\omega)$ we obtain an effective nonrecursive filter. The unit-impulse response is approximated by a sequence of finite length and the selective properties of the system are preserved. Functions $w(t)$ and $W(\omega)$ are called *time (lag) windows* and *spectral windows*, respectively.

There is a wealth of literature (Blackman and Tukey, 1958; Kurita, 1969; Kanasewich, 1973; Båth, 1974, among others) dealing with the realization of convenient window shapes (sometimes called "window carpentry"). The basic requirements are the following:

(1) $W(\omega)$ should be concentrated about the center frequency. In other words we require the main lobe of $W(\omega)$ to have small bandwidth. In general, the bandwidth must be of the same order as the width of the narrowest details of interest in $H(\omega)$. We may control the width of the main lobe by the length, N, of $w(n\Delta t)$: Narrow main lobes require large N.

(2) Side lobes of $W(\omega)$ should be small compared with the main lobe. Large side lobes may cause large contributions to $H(\omega)$ from frequency ranges distant from

the center frequency, a phenomenon called *leakage* (see e.g. Koopmans, 1974, p. 185). The amplitude of side lobes may be influenced by the type of truncation, or *tapering*, of the time window. Smooth termination usually reduces side lobes.

Properly selected windows smooth out sharp changes in $H(z)$ but otherwise preserve the characteristics of the filter. Simple time windows like *Hamming's* (Blackman and Tukey, 1958):

$$w_{1n} = \begin{cases} 0.54 + 0.46 \cos(\pi n/N) & \text{for } |n| = 0, 1, 2, \ldots, N \\ 0 & \text{for } |n| > N \end{cases}$$

or *von Hann's* (Blackman and Tukey, 1958):

$$w_{2n} = \begin{cases} 0.50 + 0.50 \cos(\pi n/N) & \text{for } |n| = 0, 1, 2, \ldots, N \\ 0 & \text{for } |n| > N \end{cases}$$

usually satisfy the requirements mentioned above. Applications of these two windows are commonly called *hamming* and *hanning*. Hamming gives smaller side lobes when compared with hanning. On the other hand, w_{2n} falls off more rapidly than w_{1n} does. The usage of spectral weighting sequence (0.25, 0.5, 0.25) is sometimes called *Hanning filtering*. Robinson (1972) proposed a window which preserves the response through the first side lobe. Within the second side lobe a function linearly decreasing from 1.0 to 0 is applied. Various types of weighting functions applied to nonrecursive filtering may be found e.g. in Kaiser (1966) or elsewhere.

A general conclusion following from our discussion is that nonrecursive filters which give good approximations to the corresponding analog form require long impulse responses, thus making the processing rather inefficient. Recursive filters, described in the following section, provide reasonable results with computational efficiency.

2.2 RECURSIVE FILTERING

Let us suppose that a digital filter is given by its system function (not by the approximation $\tilde{H}(z)$) defined in [2.1]. Further, let $x = (x_0, x_1, \ldots, x_p)$ be a one-sided finite-length input sequence. The z-transform of the output sequence, which in general will be of infinite length, is:

$$Y(z) = H(z) X(z) = \frac{\sum_{n=0}^{M} g_n z^{-n}}{\sum_{n=0}^{L} f_n z^{-n}} \sum_{n=0}^{P} x_n z^{-n}$$

Using the power-series expression for $Y(z)$ and multiplying both sides

RECURSIVE FILTERING

by $\sum_{n=0}^{L} f_n z^{-n}$ we have:

$$\sum_{n=0}^{\infty} y_n z^{-n} \sum_{n=0}^{L} f_n z^{-n} = \sum_{n=0}^{M} g_n z^{-n} \sum_{n=0}^{P} x_n z^{-n} \qquad [2.5]$$

In general, [2.5] holds for any length of the input and output sequencies and for any degree of polynomials of $H(z)$. Thus, any particular term of the output sequence may be determined by comparing coefficients of like powers of z from both sides of [2.5]. For example, for the first two terms we have:

$y_0 = x_0 g_0/f_0$ and $y_1 = (x_1 g_0 + x_0 g_1 - y_0 f_1)/f_0$

and the nth output value is:

$$y_n = \frac{1}{f_0} \sum_{i=0}^{M} g_i x_{n-i} - \frac{1}{f_0} \sum_{i=1}^{L} f_i y_{n-i} \qquad [2.6]$$

For practical use [2.6] is usually simplified with respect to the constant $1/f_0$. The system function in [2.1] remains unchanged if we multiply both the numerator and denominator by a constant $1/f_0$, provided that the first denominator coefficient $f_0 \neq 0$. Then, [2.1] may be rewritten in a form:

$$H(z) = \frac{Y(z)}{X(z)} = \frac{a_0 + a_1 z^{-1} + a_2 z^{-2} + \ldots + a_M z^{-M}}{1 + b_1 z^{-1} + b_2 z^{-2} + \ldots + b_L z^{-L}} \qquad [2.7]$$

where $a_i = g_i/f_0$ for $i = 0, 1, 2, \ldots, M$ and $b_j = f_j/f_c$ for $j = 1, 2, \ldots, L$. The nth output value becomes:

$$y_n = \sum_{i=0}^{M} a_i x_{n-i} - \sum_{i=1}^{L} b_i y_{n-i} \qquad [2.8]$$

where y_n is assumed to be zero for $n < 0$. Equations [2.6] and [2.8] describe the recursive filtering procedure. They provide computational algorithms that make it possible to generate the output sequence for successive values of n. From [2.8] we may easily obtain a set of $n+1$ equations defining the $n+1$ filtered output terms. Note that, in general, the output values are determined in terms of the present and past input and the past output. A digital filter is completely defined by the sequences a and b (or g and f), so that designing a digital filter means finding the constants a_i and b_j which satisfy the given filtering requirements.

A check of [2.8] reveals that the output sequence y is given in terms of two convolution summations. While the first right-hand summation is a convolution of the numerator of $H(z)$ and the input, the second summation is generally a convolution of the denominator of $H(z)$ and the output delayed one time-unit. Since a portion of the output is recirculated back and subtracted from the input, the recursive filter described by [2.8] may be considered as a feedback system. Making use of the z-transform representation we have:

$$Y(z) = H(z) \, X(z) = \frac{a_0 + a_1 z^{-1} + \ldots + a_M z^{-M}}{1 + b_1 z^{-1} + \ldots + b_L z^{-L}} X(z)$$

or:

$$Y(z) = (a_0 + a_1 z^{-1} + \ldots + a_M z^{-M}) \, X(z) - (b_1 + b_2 z^{-1} + \ldots \\ + b_L z^{-L+1}) \, z^{-1} \, Y(z)$$

The principles of convolution and recursive filtering are shown in Fig. 2.1.

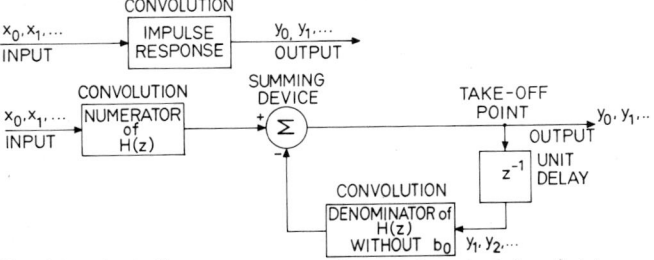

Fig. 2.1. Block-diagram representations of the principles of: (a) convolution filtering; (b) recursive filtering. The box with z^{-1} inside denotes a block producing an input–output delay equal to the sampling period Δt.

2.2.1 First-order recursive filtering

Depending on the degree of the denominator in [2.7] recursive filters of various orders may be designed. For example, a simplified form of a first-order recursive filter is:

$$y_n = ax_n - by_{n-1} \qquad [2.9]$$

Note, that the filter uses one input and one previous output value only. Here again we shall assume that causal property so that no output occurs before x_0 is applied to the input. In other words $y_n = 0$ for $n < 0$. Making use of [2.9] the successive output values are:

$$y_0 = ax_0$$
$$y_1 = ax_1 - by_0 = ax_1 - abx_0$$
$$y_2 = ax_2 - by_1 = ax_2 - abx_1 + ab^2 x_0$$
.
.
.
$$y_n = ax_n - by_{n-1} = ax_n - abx_{n-1} + \ldots + (-1)^n ab^n x_0 \qquad [2.10]$$

It follows from this set of equations that for this filter to be stable the feedback term must satisfy the inequality:

$$-1 < b < 1$$

RECURSIVE FILTERING

Evidently, for $|b|>1$ the fraction of the output which is recirculated back to the input will increase without limit as the index n increases. Consequently, the output sequence y becomes unbounded even for a bounded input.

Equation [2.9] shows that the system function in this simple case is a ratio with a one-term numerator and a two-term denominator. Applying the z-transform to both sides of [2.9] we have:

$$Y(z) = aX(z) - bz^{-1} Y(z)$$

and the system function becomes:

$$H_1(z) = \frac{Y(z)}{X(z)} = \frac{a}{1+bz^{-1}}$$

Many authors define the system function in terms of the variable z instead of z^{-1}. It is obvious that $H_1(z)$ will not change if we use the alternate form:

$$H_2(z) = \frac{az}{z+b}$$

obtained by multiplying numerator and denominator by z. Since the system is the same, the successive output coefficients are again given by the set of equations [2.10] and the same stability criterion, $|b|<1$, holds. Nevertheless, one distinction has to be emphasized: whereas the function $H_1(z)$ has a pole at $\beta_1 = -1/b$ in the z^{-1}-plane the function $H_2(z)$ has a pole at $B_1 = -b$ in the z-plane. For a stable system this means that in the former case the pole is located outside while in the latter case the pole is located inside the unit circle. For both functions $H_1(z)$ and $H_2(z)$ the unit circle divides the z^{-1}-or z-plane into stable and unstable regions. As has been discussed in Section 1.8 (see also Fig. 1.7) the criterion of the location of poles inside and outside the unit circle in the z^{-1}- and z-plane, respectively, has a general validity for all stable systems. Some confusion may arise when pole configurations from various books are compared.

By setting $z = e^{j\omega\Delta t}$, i.e. $|z|=1$, we have for the first-order system:

$$H(z) = H(e^{j\omega\Delta t}) = \frac{a}{1 + be^{-j\omega\Delta t}} = |H(\omega)| e^{j\phi(\omega)} \qquad [2.11]$$

which is the complex response function of the filter. By moving the complex variable z along the unit circle in the z-plane, or in other words by varying the angular frequency ω in the interval of interest, the amplitude and phase response of the filter are determined.

To obtain the system function in a polynomial form and the unit-impulse response, we perform the long division $a/(1 + bz^{-1})$ or determine the output sequence corresponding to the input impulse $x = (1, 0, 0, \ldots)$. The response to the input sequence x is the impulse response, and its z transform is $H(z)$. From the set of equations [2.10] and x as the input we obtain:

$$y = (a,\ -ab,\ ab^2,\ -ab^3,\ ab^4,\ \ldots)$$

Introducing numerical values, e.g. $a = 1$, $b = 0 \cdot 90$, the unit-impulse response of the filter becomes:

$$h = (1,\ -0.900,\ 0.810,\ -0.729,\ 0.656,\ \ldots)$$

The numerical example illustrates well the efficiency of recursive filters when compared with convolution filters. Applying a recursive first-order filter defined in [2.9] the determination of any output term requires one multiplication and one addition. Provided that $a = 1$, $b = 0.90$ the decay of the sequence h is rather slow and we need 21 terms before their value decreases beneath 0.1. It means that when the convolution approach is used the computer has to carry out about twenty multiplications and additions in order to determine each output value with a reasonable accuracy. In the case presented the recursive filter requires less computing time by a factor of about twenty. In our case, the decay of the unit-impulse response is controlled by the value of the coefficient b. As b approaches unity, the decay of the sequence h becomes slower. Shanks (1967) e.g. presents a first-order recursive filter with coefficients $a = 1.0$ and $b = -0.95$. The unit-impulse response for this filter requires about 45 coefficients before it decays to 10% of the h_0-value. The block-diagram representation of a first-order recursive filter may be obtained from Fig. 2.1b after introducing proper values for the numerator and denominator of $H(z)$.

2.2.2 Second-order recursive filtering

Assuming the first two denominator coefficients in [2.7] to be nonzero, we obtain the system function for a second-order recursive filter as:

$$H(z) = \frac{a_0}{1 + b_1 z^{-1} + b_2 z^{-2}} \qquad [2.12]$$

where for simplicity all numerator coefficients, except a_0, are assumed to be equal to zero. Similarly to the previous case, the z-transform of the output sequence is:

$$Y(z) = a_0 X(z) - b_1 z^{-1} Y(z) - b_2 z^{-2} Y(z)$$

Equating the terms of like powers of z yields the input–output relationship:

$$y_n = a_0 x_n - b_1 y_{n-1} - b_2 y_{n-2} \qquad [2.13]$$

Successive output values are determined in terms of the present input and the two most recent output values. As follows from [2.13], the corresponding block-diagram representation requires two delay elements. The *filter realization* derived directly from [2.13] is called the *direct form* and is diagrammed in Fig. 2.2a.

Alternative block-diagram forms may be found by re-arranging the expression for $H(z)$. In [2.12] the system function is given as a ratio of a constant and a second-

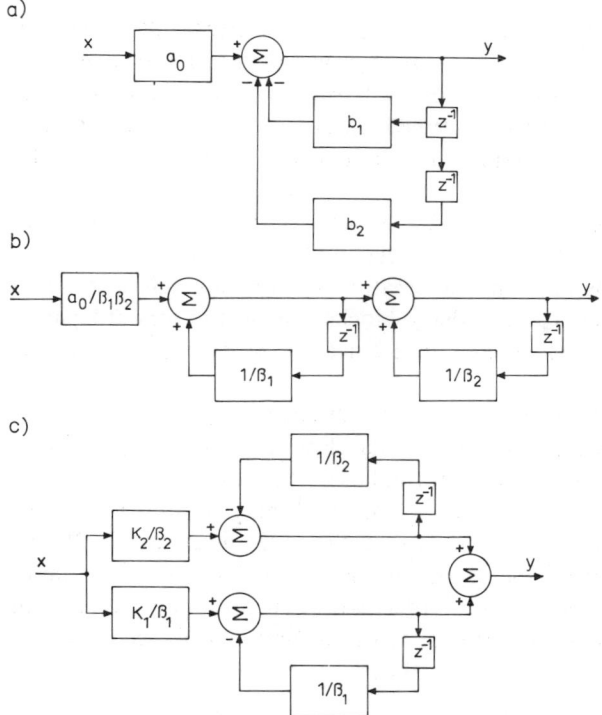

Fig. 2.2. Three block-diagram representations of a simplified second-order recursive filter: (a) direct form; (b) serial or cascade form; (c) parallel form.

degree polynomial in z^{-1}. Thus, the rational function $H(z)$ may be expressed in two alternative forms: as a product in a factored form or as a partial-fraction expansion. This means that the system function may be written as:

$$H(z) = \frac{a_0}{(z^{-1} - \beta_1)(z^{-1} - \beta_2)} \qquad [2.14]$$

or as: $$H(z) = \frac{K_1}{z^{-1} - \beta_1} + \frac{K_2}{z^{-1} - \beta_2} \qquad [2.15]$$

where β_1 and β_2 are the poles of $H(z)$, corresponding to the roots of the denominator. Assuming, for example, that $H(z)$ contains two real first-order poles, constants K_1 and K_2 are easily determined as:

$$K_i = [(z^{-1} - \beta_i) H(z)]_{z^{-1} = \beta_i} \text{ where } i = 1, 2 \qquad [2.16]$$

The right-hand side of [2.14] is a product of system functions of two simple first-order filters. Thus, $H(z)$ may be viewed as the frequency response of a *serial*, or *cascade*, *arrangement* of two first-order subsystems, where the output of the first subsystem is applied to the input of the second one. The corresponding network is shown in Fig. 2.2b. Similarly, [2.15] may be interpreted in terms of a *parallel arrangement* of two first-order subsystems. Both have a common input and the sum of the two outputs provides the final output sequence y (see Fig. 2.2c). Various stages in constructing the block diagrams are discussed in more detail, e.g. by Ackroyd (1973).

In the block diagrams presented in Fig. 2.2, several simplifications have been made. Firstly, the numerator of $H(z)$ has been substituted by a constant a_0. This is of course a special case. Generally, the numerator is a polynomial in z^{-1} and consequently the block diagrams contain delay blocks also in the forward paths. Note that the three forms represented in Fig. 2.2 use all the delay blocks in the feedback paths.

Secondly, only real first-order poles of $H(z)$ were discussed. However, $H(z)$ may contain also multiple and complex-conjugate poles. For multiple poles, [2.16] is replaced by a somewhat more complicated formula (see Section 2.5.1). In case of complex-conjugate poles, K_1 and K_2 are also complex conjugate and we do not carry out the polynomial factorization but utilize the form given in [2.12]. An important conclusion follows from the foregoing discussion: assuming real input and output values, a recursive filter of any order, with real or complex-conjugate poles, may be substituted by serial arrangements of first- and/or second-order filters.

2.2.3 Lth-order recursive filtering

The most general form of a system function of an Lth-order recursive filter is described by [2.7] where the coefficient $b_L \neq 0$. Performing cross-multiplication in [2.7] we find:

$$(1 + b_1 z^{-1} + b_2 z^{-2} + \ldots + b_L z^{-L}) Y(z) = (a_0 + a_1 z^{-1} + a_2 z^{-2} + \ldots + a_M z^{-M}) X(z) \qquad [2.17]$$

The multiplication in [2.17] and inverse z-transform yield the recursion formula for the output sequence:

$$y_n = a_0 x_n + a_1 x_{n-1} + \ldots + a_{M-1} x_{n-M+1} + a_M x_{n-M} - b_1 y_{n-1} - b_2 y_{n-2} - \ldots - b_{L-1} y_{n-L+1} - b_L y_{n-L}$$
$$= \sum_{i=0}^{M} a_i x_{n-i} - \sum_{i=1}^{L} b_i y_{n-i}$$

which is identical with [2.8]. As follows from the formula, the present output term depends upon the present input value, M most recent input values and L most recent output values. The direct form of the block-diagram arrangement is depicted in Fig. 2.3a. The left half of the network contains M delay elements in the forward

RECURSIVE FILTERING

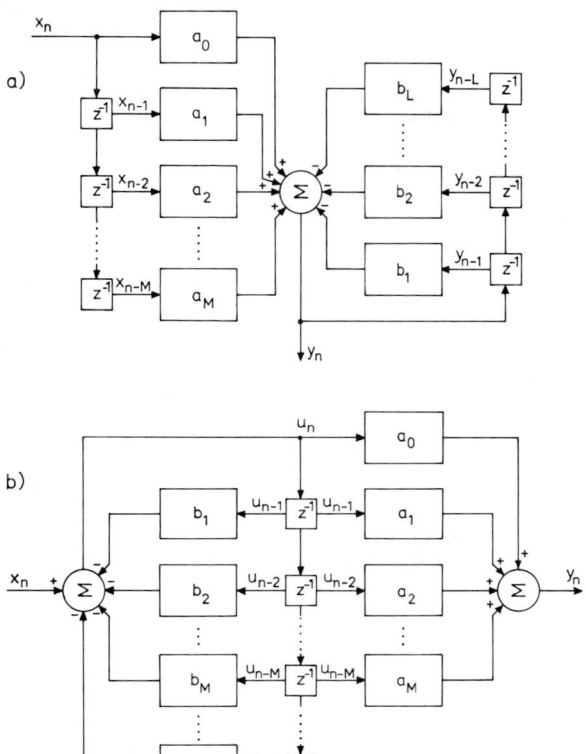

Fig. 2.3. Block-diagram representation for an Lth-order recursive filter: (a) direct form; (b) canonic form, where $L > M$.

paths corresponding to the numerator of $H(z)$. The right half contains L delay blocks corresponding to the Lth-order of the filter.

An alternate, so-called *canonic form* of the block-diagram representation may be derived from [2.7] written in the following way:

$$H(z) = \frac{Y(z)}{X(z)} = \frac{\sum_{i=0}^{M} a_i z^{-i}}{1 + \sum_{i=1}^{L} b_i z^{-i}} \qquad [2.18]$$

The z-transform of the output is:

$$Y(z) = \frac{X(z)}{1 + \sum_{i=1}^{L} b_i z^{-i}} \sum_{i=0}^{M} a_i z^{-i} = U(z) \sum_{i=0}^{M} a_i z^{-i} \qquad [2.19]$$

where:

$$U(z) = \frac{X(z)}{1 + \sum_{i=1}^{L} b_i z^{-i}} \qquad [2.20]$$

It is evident that a network which would perform operations according to [2.18], [2.19] and [2.20] could be considered as a combination of two arrangements. The first one is a feedback network with the sequence x as an input and u as an output, as described by [2.20]. The second network contains only forward paths and uses the sequence u as an input according to [2.19]. Inverse z-transforms of $U(z)$ and $Y(z)$ are:

$$u_n = x_n - \sum_{i=1}^{L} b_i u_{n-i} \quad \text{and} \quad y_n = \sum_{i=0}^{M} a_i u_{n-i}$$

respectively. Since the sequence u is applied as an input into the second network, the total block diagram can be constructed in canonic form (Fig. 2.3b). The essential property of the canonic form is that the minimum number of delay units is used in the realization. Note that the same delay blocks are used in both the forward and feedback paths, i.e. for both the zeros and poles of $H(z)$.

Serial and parallel forms may be constructed via factorization and partial-fraction expansion of the system function respectively, regardless of the order of the filter. As mentioned above, both forms may be constructed by means of simple first- and second-order filters only. This procedure naturally requires the knowledge of zeros and poles of $H(z)$. To find the roots of high-degree polynomials in both the numerator and denominator of $H(z)$ may create certain additional problems.

2.3 SYSTEM FUNCTIONS OF RECURSIVE FILTERS

In Section 1.4 we discussed the evaluation of the frequency-response function of an analog system by making use of a harmonic input signal. A similar approach may be followed in the case of digital systems. Consider a causal system with a pulse-transfer function $H(z) = \mathcal{Z}\{h_n\}$ and a stationary input sequence:

$$x_n = \sin \Omega n = (1/2j)\left[e^{j\Omega n} - e^{-j\Omega n}\right]$$

Let us emphasize that digitization has here been carried out with a sampling period $\Delta t = 1$ sec. According to the convolution summation [1.42] the nth term of the output sequence is:

$$y_n = (1/2j) \sum_{i=0}^{\infty} \left[e^{j\Omega(n-i)} - e^{-j\Omega(n+i)}\right] h_i$$

By rearranging the right-hand side and substituting $z = e^{j\Omega}$ we have:

SYSTEM FUNCTIONS OF RECURSIVE FILTERS

$$y_n = (1/2j) [e^{j\Omega n} - e^{-j\Omega n}] \sum_{i=0}^{\infty} z^{-i} h_i = H(e^{j\Omega}) \sin \Omega n \qquad [2.21]$$

Thus, the output of the digital system, which has been excited by a digitized sinusoidal signal, gives the value of the frequency response function at z corresponding to $\omega = \Omega$ on the unit circle. If we let the angular frequency vary in the entire frequency range of interest, in other words, if we let z move along the unit circle in the complex z-plane, we determine the system function from [2.21] as a function of frequency.

An extremely important behaviour of the system function follows immediately from [2.21]. Since:

$$H(e^{j\Omega}) = H(e^{j\Omega \pm j2\pi n}) \quad \text{for } n = 0, 1, 2, \ldots$$

the system function of any digital filter is a periodic function of frequency with a period 2π, assuming a unit sampling period. For an arbitrary Δt we find a similar relation:

$$H(e^{j\Omega \Delta t}) = H(e^{j\Omega \Delta t \pm j2\pi n}) = H\left[\exp\left\{j\Delta t \left(\Omega \pm \frac{2\pi n}{\Delta t}\right)\right\}\right] \quad \text{for } n = 0, 1, 2, \ldots$$

Generally, $H(e^{j\Omega \Delta t})$ repeats itself along the Ω-axis with a period of $2\pi/\Delta t$. In order to determine the amplitude and phase response, the complex function $H(z)$ is written in a polar form. For $\omega = \Omega$ and $\Delta t = 1$ sec we have:

$$H(z) = H(e^{j\Omega}) = M(\Omega) \, e^{j\phi(\Omega)} = \left[P^2(\Omega) + Q^2(\Omega)\right]^{\frac{1}{2}} \exp\left[\arctan \frac{Q(\Omega)}{P(\Omega)}\right] \qquad [2.22]$$

where $P(\Omega) = \text{Re } H(e^{j\Omega})$ and $Q(\Omega) = \text{Im } H(e^{j\Omega})$.

As an example consider the simple system discussed in Section 1.7 with the polynomial system function:

$$H(z) = 2 + z^{-1}$$

In order to determine the value of $H(z)$ for an arbitrary angular frequency, $\omega = \Omega$, we let the variable z move along the unit circle in the z-plane into the point $z = e^{j\Omega}$. Introduction of $z^{-1} = e^{-j\Omega}$ into $H(z)$ provides the frequency response for $\omega = \Omega$. For example for $\omega = 0$ the variable $z^{-1} = 1$ and $H(z) = H(1) = 2 + 1 = 3$. Applying the polar form we have $M(0) = 3$ and $\phi(0) = 0$. Numerical values are easily available for the following angular frequencies:

ω	$H(e^{j\omega})$	$M(\omega)$	$\phi(\omega)$
0	3	3	0
$\pi/2$	$2 - j$	$\sqrt{5}$	$\arctan(-1/2) \doteq -27°$
π	1	1	0
$3\pi/2$	$2 + j$	$\sqrt{5}$	$\arctan(1/2) \doteq 27°$
2π	3	3	0

The amplitude and phase responses are plotted in Fig. 2.4 as a function of ω. The response is not very good since it distorts the input spectrum at all frequencies between 0 and the folding frequency $\omega_N = \pi$. The periodicity of the frequency response, with a period of 2π, is obvious from Fig. 2.4.

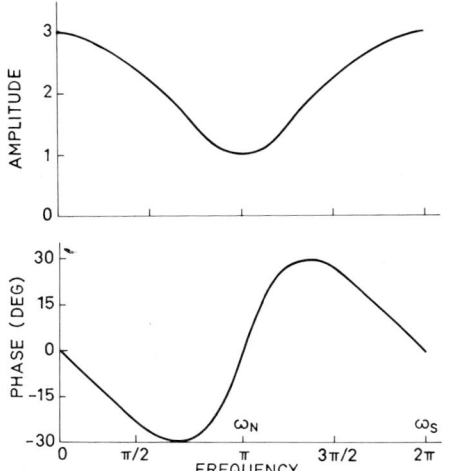

Fig. 2.4. Amplitude (arbitrary units) and phase responses of the system $H(z) = 2 + z^{-1}$.

It is true that [2.21] and [2.22] define the frequency characteristics. Nevertheless they are hardly applicable in a filter design. The real- and imaginary-part separations shown in the example above are usually rather laborious even for simple systems. Further and more importantly, the frequency selectivity of the response function is not immediately visible from either of these two definitions. This of course complicates the design. In the following section we shall discuss the so-called pole-zero technique which in general is an effective method with a direct physical interpretation.

2.4 POLE-ZERO TECHNIQUE

In Sections 2.1 and 2.2 we have considered system functions which were either polynomials or rational fractions of polynomials in z^{-1}. In the second case the numerator as well as the denominator of $H(z)$ may be factored and $H(z)$ may be uniquely determined, except for a constant multiplier, by its zeros and poles for any value of z^{-1}. In the first case $H(z)$ is defined solely by its zeros, since it does not contain any poles. The advantages of the *pole-zero technique* may be illustrated by using several simple systems. For example, the system function $H(z) = 2 + z^{-1}$ has one zero for $\alpha_1 = -2$. The location of this zero with respect to the unit circle in the z^{-1}-plane is illustrated in Fig. 2.5a. Hereafter we shall indicate the location of zeros and poles in the z- or z^{-1}-plane by circles and crosses, respectively. Point C

POLE-ZERO TECHNIQUE

on the unit circle, corresponding to the angular frequency Ω and sampling period $\Delta t = 1$ sec, shows the complex value $z^{-1} = e^{-j\Omega}$. As follows from the figure, the amplitude response for $\omega = \Omega$ is given by the distance between the point C and the zero α_1. Values of $M(\omega)$ for other frequencies are determined by letting z^{-1} rotate about the unit circle and by measuring the corresponding distances $\overline{C\alpha_1}$. The associated phase shift is measured by the angle θ so that $\phi(\omega) = \theta(\omega)$.

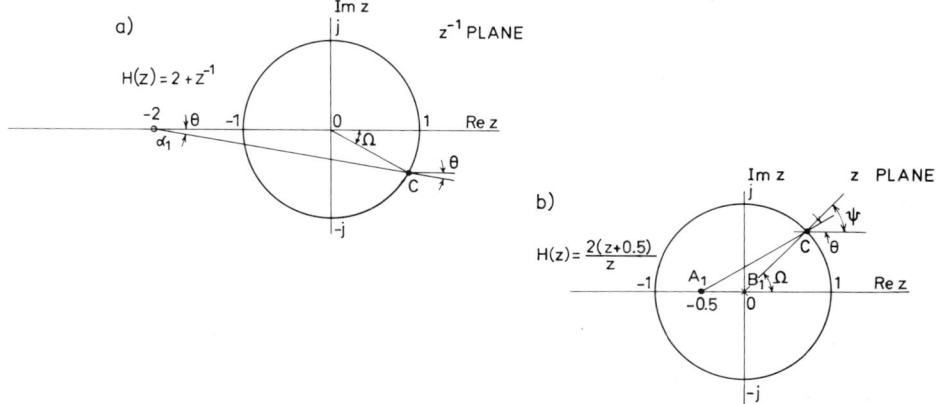

Fig. 2.5. Geometrical representation of the system function $H(z) = 2 + z^{-1} = 2(z + 0.5)/z$: (a) in the z^{-1}-plane; (b) in the z-plane.

Some authors prefer to work with positive powers of z. Our system function may also be written as:

$$H(z) = 2 + z^{-1} = 2(z + 0.5)/z$$

$H(z)$ expressed in this form has one first-order zero at $A_1 = -0.5$ and one first-order pole at $B_1 = 0$. Since the system function remains the same, the pole-zero technique must provide the same results for both forms of $H(z)$. Point C on the unit circle now depicts the complex variable $z = e^{j\Omega}$. From Fig. 2.5b it is seen that the amplitude and phase responses for $\omega = \Omega$ are:

$$M(\Omega) = 2\overline{CA_1}/\overline{CB_1} \text{ and } \phi(\omega) = \theta - \psi$$

Values for other angular frequencies are obtained by moving the variable z along the unit circle. Introducing numerical values, e.g. $\omega = 0, \pi/2, \pi, 3\pi/2, 2\pi$, we obtain results identical with those calculated from [2.22].

For those familiar with analog control-system design by means of the Laplace transform, the parallel is obvious: whereas for analog systems the frequency is measured on the imaginary axis in the complex p-plane, for digital systems we measure the frequency along the unit circle in the complex z^{-1}- or z-plane. The unit circle in the z^{-1}- or z-plane is of the same importance as is the $j\omega$-axis in the p-plane, since it separates the stable and unstable regions.

Making use of the graphical pole-zero representation, we may directly follow the development of the frequency characteristics as a function of the angular frequency. For example, Fig. 2.5a shows, that the gain of $H(z)$, given by $\overline{C\alpha_1}$, will reach its maximum value for $\omega = 0$. As the angular frequency increases, the gain decreases and is minimal for the folding frequency $\omega = \omega_N = \pi$. Due to the periodicity, the gain function is symmetric around ω_N. The least gain at $\omega = \pi$ is equal to one third of the maximum gain value corresponding to zero frequency. It is possible to make the minimum gain equal to zero simply by shifting the zero at $\alpha_1 = -2$ to a new position at $\alpha_2 = -1$. This new system, $H_1(z) = 1 + z^{-1}$, will completely reject components with $\omega = \omega_N$. A simple filter like $H_1(z)$ has here been used only to illustrate some of the zero-pole principles, but it would not be of much use in real applications. Besides the fact that the component $\omega = \omega_N$ is filtered out, the filter also distorts the input spectrum in the entire frequency range which in general is an undesirable behaviour.

In a way similar to that shown in the preceding case, we may construct the system function of a filter which would reject the direct component (D.C.). These filters are of special interest since they remove the very low-frequency components from the signal. Considering a system function with a first-order zero at $\alpha_1 = 1$, we have:

$$H(z) = 1 - z^{-1}$$

For the sake of simplicity we again consider only frequencies $\omega = 0, \pi/2, \pi, 3\pi/2$ and 2π. The corresponding gain values then are $M(\omega) = 0, \sqrt{2}, 2, \sqrt{2}$ and 0. The gain is zero for the zero frequency, so-called *D.C. rejection filter*, and reaches its maximum for $\omega = \pi$. The filter again distorts the input spectrum in the entire frequency range and is therefore of little practical use. An ideal D.C. rejection filter should have zero amplitude characteristics for the zero frequency and flat, nonzero characteristics for remaining frequencies. A reasonable approximation of this ideal filter may be realized by adding a first-order pole. Thus:

$$H_2(z) = \frac{H(z)}{1-bz^{-1}} = \frac{1-z^{-1}}{1-bz^{-1}}$$

The system function $H_2(z)$ has a zero at $\alpha_1 = 1$ and a pole at $\beta_1 = 1/b$. For any ω the amplitude characteristics are determined as a ratio $\overline{C\alpha_1}/\overline{C\beta_1}$ where C again is the point on the unit circle. Clearly, $M_2(\omega)$ will be zero for $\overline{C\alpha_1} = 0$ and approximately constant and close to unity for those frequencies for which $\overline{C\alpha_1} \simeq \overline{C\beta_1}$. In other words, β_1 has to be located near to α_1 to make the ratio close to unity for any ω except for ω approaching zero. Another requirement follows from the stability criterion, $|b|<1$, which means that the pole has to be located outside the unit circle. For example for $b = 0.95$ we have:

$$H_2(z) = \frac{1-z^{-1}}{1-0.95\,z^{-1}} \qquad [2.23]$$

POLE-ZERO TECHNIQUE

Amplitude characteristics normalized with respect to the maximum gain (for $\omega = \pi$) is depicted in Fig. 2.6a. As expected the filter totally rejects the D.C. component but transmits all appreciably higher frequencies with a practically constant gain. In fact, for frequencies $0.03\omega_N$ and $0.15\omega_N$ the gain is about $0.95 M_{max}$ and $0.99 M_{max}$, respectively. In other words, the filter $H_2(z)$ has the properties required for a D.C. rejection filter. The decisive factor influencing the shape of $M_2(\omega)$ is the distance, $\overline{\alpha_1\beta_1}$, between the pole and the zero. The shorter the distance $\overline{\alpha_1\beta_1}$, the narrower is the range of rejected frequencies.

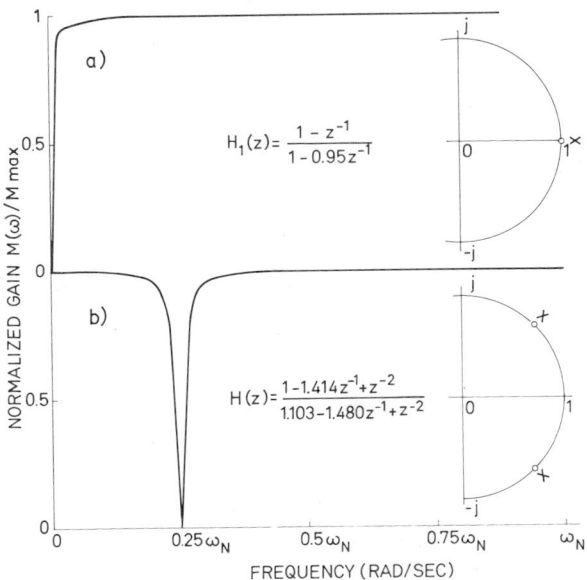

Fig. 2.6. Normalized amplitude characteristics and poles (crosses) and zeros (circles) configurations: (a) D.C. rejection filter; (b) $0.25 \omega_N$-rejection filter.

Multiplication of both sides of [2.23] by $(1 - 0.95 z^{-1}) X(z)$ gives:

$$(1 - 0.95 z^{-1}) Y(z) = (1 - z^{-1}) X(z)$$

and the recursion formula for the nth output value becomes:

$$y_n = x_n - x_{n-1} + 0.95 y_{n-1}$$

The block-diagram representation requires two delay units, one in the forward and the other in the feedback path.

Rejection filtering is not limited to D.C. components; it may be applied to rejection of any frequency within the range $0-\omega_N$. Suppose that our recorded signal contains an unwanted sinusoidal noise component with a period of 4 sec. The corresponding angular frequency $\Omega = 2\pi/4 = 0.5\pi$ rad/sec. Digitization of the trace with a sampling period $\Delta t = 0.5$ sec gives the folding frequency $\omega_N = \pi/\Delta t =$

2π rad/sec. Thus in order to reject the $\Omega=0.25\,\omega_N$ component we place a complex zero at a point on the unit circle corresponding to an angle $(\Omega/\omega_N)\,180°=45°$. Since we expect that a real input sequence will produce a real output, the system function cannot be built up by using single complex zeros and/or poles. Complex singularities of the system function must always appear as complex-conjugate pairs. Therefore, we place the second zero at a point corresponding to $-45°$. The complex-conjugate zeros are:

$$\alpha_{1,2} = \cos 45° \pm j \sin 45° = 0.707 \pm j\, 0.707$$

To improve the amplitude characteristics we locate a pair of complex-conjugate poles close to $\alpha_{1,2}$ and outside the unit circle as in the preceding case. Let us use poles:

$$\beta_{1,2} = 0.740 \pm j\, 0.740$$

which together with $\alpha_{1,2}$ give the system function:

$$H(z) = \frac{(z-\alpha_1)(z-\alpha_2)}{(z-\beta_1)(z-\beta_2)} = \frac{1-1.414\,z^{-1}+z^{-2}}{1.103-1.480\,z^{-1}+z^{-2}} \qquad [2.24]$$

The amplitude characteristics, normalized with respect to the maximum-gain value are plotted in Fig. 2.6b. While a sharp decrease of the gain for frequencies near to $0.25\,\omega_N$ is evident from the diagram, the relative gain remains constant and close to 1 for most other frequencies within the range from 0 to 1 cps. Whereas the filter completely rejects the $0.25\,\omega_N$-component the components with frequencies, say, lower than $0.2\,\omega_N$ and higher than $0.3\,\omega_N$ are transmitted with negligible amplitude distortion. Filters with these properties are called *notch filters*. As in the preceding case, the decisive factor of the sharpness of the cut-off is the pole-zero distance.

The recursion relation for this filter follows from [2.24]. Dividing both the numerator and denominator of $H(z)$ by a factor of 1.103 and by cross-multiplying we obtain:

$$(1 - 1.342\,z^{-1} + 0.907\,z^{-2})\,Y(z) = (0.907 - 1.282\,z^{-1} + 0.907\,z^{-2})\,X(z)$$

which yields:

$$y_n = 0.907\,x_n - 1.282\,x_{n-1} + 0.907\,x_{n-2} + 1.342\,y_{n-1} - 0.907\,y_{n-2}$$

Members of the output sequence y depend upon the present and two most recent input values and upon the two most recent output values. The corresponding block-diagram arrangement requires four delay units, two in the forward and two in the feedback path.

By making use of the geometrical interpretation of system-function singularities in the z- or z^{-1}-plane, the amplitude and phase characteristics can be determined. Let $\alpha_1, \alpha_2, \ldots, \alpha_M$ and $\beta_1, \beta_2, \ldots, \beta_L$ be the zeros and poles, respectively, and let

C be the point on the unit circle corresponding to the angular frequency $\omega = \Omega$. Then the amplitude characteristics become:

$$M(\Omega) = K \prod_{i=1}^{M} \overline{C\alpha_i} \Big/ \prod_{i=1}^{L} \overline{C\beta_i} \quad (K \text{ real constant}) \tag{2.25}$$

and the phase is given (see Fig. 2.5b) by:

$$\phi(\Omega) = \sum_{i=1}^{M} \theta_i - \sum_{i=1}^{L} \psi_i \tag{2.26}$$

The examples discussed above illustrate the intuitive usefulness of the pole-zero technique in designing digital systems (see also Shanks, 1967, and Mooney, 1968). Unfortunately, the complexity of the method increases rapidly with the increasing number of system-function singularities. Thus, the applicability is limited to filters of lower order. For complicated systems a more systematic design technique is required.

2.5 APPROXIMATION OF ANALOG SYSTEMS

The behaviour of a large number of filters simulated nowadays by means of digital systems has been described and frequently used in analog form long before the advent of digital computers. The literature provides a wealthy list of analog filters satisfying diverse requirements which may occur in practical applications. Naturally, it would be extremely useful if the extensive results already available for the synthesis of analog filters could be also applied in the digital-filter design. Below, we discuss several possibilities for constructing a system function $H(z)$ which significantly resembles a given analog system in a certain sense.

2.5.1 *Impulse invariance*

One possibility of approximation is to define the digital system in a way which would ensure the resemblance of the unit-impulse responses of both the analog and corresponding digital systems. In other words, we search for a digital system $H(z)$ with an impulse-response sequence h which at the sampling instants is equal to the impulse response $h(t)$ of the analog system. We wish that:

$$h_n = [h(t)]_{t=n\Delta t} \quad \text{for } n = 0, 1, 2, \ldots \tag{2.27}$$

where h_n is the nth sample of the impulse-response sequence h. This approach is called *impulse invariance* (Rader and Gold, 1967; Gold and Rader, 1969). The whole procedure is performed in the following successive steps:

(1) The transfer function $H(p)$ is given.
(2) Carry out the inverse Laplace transform in order to obtain the unit-impulse response $h(t) = \mathcal{L}^{-1}\{H(p)\}$.

(3) Digitize $h(t)$, with the sampling period Δt, to find the sequence $\{h_n\}$.
(4) z-transform of $\{h_n\}$ yields the required system function $H(z) = \mathcal{Z}\{h_n\}$. By comparison of the initial and final functions we can determine the investigated correspondence $H(p) \to H(z)$. For common types of time functions tables of the Laplace and z-transforms (see e.g. Table II) considerably simplify the whole procedure. Below we look at the approach in more detail for a general type of $H(p)$.

The transfer function, $H(p)$, of a linear analog system with constant parameters is usually given by a ratio of two polynomials in p such as:

$$H(p) = \mathcal{L}\{h(t)\} = \frac{u_M p^M + u_{M-1} p^{M-1} + \ldots + u_1 p + u_0}{v_L p^L + v_{L-1} p^{L-1} + \ldots + v_1 p + v_0} = \frac{U(p)}{V(p)} \quad \text{for } L \geqslant M$$

which is analogous to the form given in [2.1] for digital systems. The impulse invariance technique is not limited to real first-order poles of $H(p)$ and multiple and complex-conjugate poles may be considered as well. Let us now investigate three distinct cases, although various combinations of these are also possible.

$H(p)$ *contains first-order poles only.* Let us suppose that the roots of the polynomial $V(p)$ are real and distinct. Then the transfer function $H(p)$ can be expanded into a sum of partial fractions:

$$H(p) = \frac{U(p)}{V(p)} = \sum_{i=1}^{L} \frac{K_i}{(p - p_i)} \qquad [2.28]$$

where p_i are the poles of $H(p)$ and K_i are the constants given (see Båth, 1968, p. 23) by:

$$K_i = \left[\frac{U(p)}{V(p)} (p - p_i) \right]_{p = p_i}$$

An alternative formula for K_i is:

$$K_i = \frac{U(p_i)}{V'(p_i)} \qquad [2.29]$$

where $V'(p_i)$ is the derivative $(d/dp) V(p)$ evaluated for $p = p_i$. For the special case where $M = L$, it is necessary to carry out the long division prior to the partial-fraction expansion, in order that $H(p)$ be a proper fraction. From Table II we find that the inverse transform of any term on the right-hand side of [2.28] is:

$$\mathcal{L}^{-1}\left\{\frac{K_i}{p - p_i}\right\} = K_i \exp(p_i t)$$

Thus, the inverse Laplace transform of both sides of [2.28] yields the unit-impulse response:

$$h(t) = \mathcal{L}^{-1}\{H(p)\} = \sum_{i=1}^{L} K_i \exp(p_i t) \qquad [2.30]$$

Other methods for determining the response [2.30] may be employed as well. For example, Longman and Sharir (1971) outlined a procedure for expressing $h(t)$ by means of an infinite series. The procedure is carried out without finding the roots, p_i, in [2.28].

It is worth noting that [2.30] provides the general stability criterion directly by means of the transfer function $H(p)$. In Section 1.2.1 we mentioned that the absolute integrability of $h(t)$ assures the stability of the system. Thus, for a stable system all terms, $K_i \exp(p_i t)$ for $i = 1, 2, \ldots, L$, in the summation in [2.30] must decay with increasing time. Evidently, this condition is satisfied only when the real parts of p_i for all i's are negative. Broadly speaking, an analog system is stable if all the poles of the transfer function lie in the left half of the complex p-plane.

Introducing discrete time instants into [2.30] we obtain a sequence:

$$h_n = \sum_{i=1}^{L} K_i \exp(p_i n \Delta t)$$

which provides the unit-impulse response of the digital impulse-invariant system. The z-transform of an exponential function has been derived in Section 1.6 and the transform of $\{h_n\}$ can be written as:

$$\mathcal{Z}\{h_n\} = H(z) = \sum_{i=1}^{L} \frac{K_i}{1-\exp(p_i \Delta t) z^{-1}} = \sum_{i=1}^{L} K_i \frac{z}{z-\exp(p_i \Delta t)}$$

where we assume that $\text{Re } p_i < 0$ for $i = 1, 2, \ldots, L$. In view of our discussion concerning impulse invariance, it is clear that the essential requirement defined in [2.27] is fulfilled via the correspondence $H(p) \to H(z)$ or:

$$\sum_{i=1}^{L} \frac{K_i}{(p-p_i)} \to \sum_{i=1}^{L} \frac{K_i}{1-\exp(p_i \Delta t) z^{-1}}$$

Note that the constants K_i and p_i defining the digital filter $H(z)$ are determined directly from $H(p)$.

It should be emphasized that the impulse-invariance method preserves the location of poles in the p- and z-planes but neglects the location of zeros. Obviously, this may affect the mentioned correspondence $H(p) \to H(z)$. There is an important category of analog filters (see Chapter 3) with the system function of a form:

$$H(p) = \frac{1}{\prod_{i=1}^{N}(p-p_i)}$$

The corresponding pulse-transfer function, $H(z)$, derived by means of the partial-fraction expansion will in general contain finite zeros in spite of the fact that there were no finite zeros in $H(p)$. However, for many practical applications the influence of these zeros may be minimized, e.g. by using a sufficiently small sampling period. More details may be found in Rader and Gold (1967).

H(p) contains a pair of complex-conjugate poles. In this case the denominator of $H(p)$ may be written as:

$$V(p) = (p - p_1)(p - p_1^*)$$

where $p_1 = -\sigma_1 + j\omega_1$ and $p_1^* = -\sigma_1 - j\omega_1$. Then the corresponding coefficients K_1 and K_1^* are also complex conjugate, so that the transfer function becomes:

$$H(p) = \frac{U(p)}{V(p)} = \frac{K_1}{(p-p_1)} + \frac{K_1^*}{(p-p_1^*)} = \frac{c+jd}{p+\sigma_1-j\omega_1} + \frac{c-jd}{p+\sigma_1+j\omega_1} \quad [2.31]$$

Making use of [2.29] and separating the real and imaginary parts we determine the values of c and d as:

$$c + jd = \frac{U(p_1)}{V'(p_1)}$$

The transfer function can be expressed in a slightly modified form:

$$H(p) = \frac{2c(p+\sigma_1) - 2d\omega_1}{(p+\sigma_1)^2 + \omega_1^2}$$

which may be found in Table II. Thus:

$$h(t) = \mathcal{L}^{-1}\{H(p)\} = 2c\exp(-\sigma_1 t)\cos\omega_1 t - 2d\exp(-\sigma_1 t)\sin\omega_1 t$$

and using Table II we find the digital system function directly in the following form:

$$H(z) = \sum_{n=0}^{\infty} h_n z^{-n}$$

$$= \frac{2c[1 - \exp(-\sigma_1\Delta t)\cos(\omega_1\Delta t)z^{-1}] - 2d\exp(-\sigma_1\Delta t)\sin(\omega_1\Delta t)z^{-1}}{1 - 2\exp(-\sigma_1\Delta t)\cos(\omega_1\Delta t)z^{-1} + \exp(-2\sigma_1\Delta t)z^{-2}}$$

Thus, the constants c, d, σ_1 and ω_1 defining the filter $H(z)$ are obtained directly from the known analog version of the system.

H(p) contains multiple poles. Let p_1 be the multiple Nth-order root of $V(p)$. Performing the partial-fraction expansion of $H(p)$, we have:

$$H(p) = \frac{K_1}{(p-p_1)^N} + \frac{K_2}{(p-p_1)^{N-1}} + \cdots + \frac{K_{N-1}}{(p-p_1)^2} + \frac{K_N}{(p-p_1)} \quad [2.32]$$

where the real values of K_i for $i = 1, 2, \ldots, N$ are:

$$K_i = \frac{1}{(i-1)!}\left[\frac{d^{i-1}}{dp^{i-1}}(p-p_1)^N H(p)\right]_{p=p_1}$$

Comparing the individual terms on the right-hand side of [2.32] with transforms listed in Table II, the correspondence:

$$\frac{K_i}{(p-p_1)^i} \rightarrow \frac{(-1)^{i-1}}{(i-1)!}\frac{d^{i-1}}{dp_1^{i-1}}\left[\frac{K_i}{1-\exp(p_1\Delta t)z^{-1}}\right]$$

APPROXIMATION OF ANALOG SYSTEMS

is obtained. As in the two preceding cases constants K_i and p_1 are evaluated from the given analog transfer function $H(p)$.

Summarizing these results, we can say that the impulse-invariance technique may be applied without respect to degree of polynomials in $H(p)$. The transfer function may contain single, multiple, real or complex poles or any combination of these. Properly combining the basic cases discussed above, the corresponding digital system can be defined. Nevertheless, some of the disadvantages of the technique should be mentioned. Firstly, since the response sequence h is the digitized form of the function $h(t)$ we would also expect high similarity between the frequency characteristics $H(p)$ and $H(z)$. However, the term high similarity has here a special meaninlg. Due to the aliasing effect, the discrepancy between the two characteristics generaly increases with frequency approaching the folding point. In other words, good resemblance can be expected only for frequencies which are rather low when compared with ω_N. Another inconvenience connected with the impulse-invariance technique is the gain dependence upon the sampling period Δt which may cause overflow difficulties during practical computations.

2.5.2 Convolution approximation

Consider an analog system $H(p)$, satisfying certain filtering requirements, an arbitrary input $x(t)$ and the resulting output $g(t)$. Let us search for a digital system which when excited by the sequence:

$$x = [x(0), x(\Delta t), x(2\Delta t), \ldots]$$

responds by a sequence which reasonably approximates $g(t)$ at the sampling instants. We denote the approximate digital system by $\tilde{H}_i(z)$ where i denotes the type of approximation used. The impulse invariance approach, discussed in the preceding section, may be viewed as a special case of the more general *output-invariance technique*. While in the former case we compare the analog and digital unit-impulse responses, in the latter case we compare responses to an arbitrary input signal. Functions $x(t)$, $g(t)$ and $h(t)$ are analog functions of time and their amplitudes may therefore be measured at any time instant, e.g. at $t = 0, \Delta t, 2\Delta t, \ldots, N\Delta t$. Corresponding sampled values then become $x(n\Delta t)$, $g(n\Delta t)$ and $h(n\Delta t)$ where $n = 0, 1, 2, \ldots, N$. By contrast, the digital system processes digital input and output signals. When the system $\tilde{H}_i(z)$ is excited by the sequence x, it produces output samples, say, y_n, whose values will differ in general from those of $g(n\Delta t)$. Apart from the aliasing effect, this difference may be attributed to the fact that while the continuous input activates the system $H(p)$ without interruption, the digital input has nonzero values only at discrete time instants. In between these instants, the system receives no information. Our effort now will be to design the system $\tilde{H}_i(z)$ in a way which would minimize the difference $y_n - g(n\Delta t)$. We shall follow an approach presented by Vích (1968).

The input–output relationship for an analog system is defined by the convolution integral as has been shown in [1.6]. Consider a one-sided input, $x(t)$, with a distinct origin or arrival time at $t = 0$, (i.e. $x(t) = 0$ for $t \leq 0$) exciting a causal system with $h(t) = 0$ for $t < 0$. Equation [1.6] remains valid if we replace the integration limits by 0 and t. Thus, the output becomes:

$$g(t) = \int_0^t x(\tau) h(t - \tau) \, d\tau$$

In order to determine the amplitudes $g(n\Delta t)$, we have to evaluate the convolution integral in the time interval $0 \leq \tau \leq n\Delta t$. Assuming equally spaced sampling with sampling interval Δt, we may apply the *Gregory formula* (Hamming, 1962, p. 140):

$$\int_0^{n\Delta t} u(x) \, dx = \Delta t \left[\sum_{k=0}^{n} u_k - \frac{1}{2}(u_0 + u_n) + \frac{1}{12}(\Delta u_0 - \Delta u_{n-1}) \right.$$
$$\left. - \frac{1}{24}(\Delta^2 u_0 + \Delta^2 u_{n-2}) + \frac{19}{720}(\Delta^3 u_0 - \Delta^3 u_{n-3}) - \cdots \right] \quad [2.33]$$

where the Δ operator is the forward difference operator so that:

$$\Delta u_n = u_{n+1} - u_n$$

and for repeated differences we have:

$$\Delta^r u_n = \Delta[\Delta^{r-1} u_n] = \Delta^{r-1} u_{n+1} - \Delta^{r-1} u_n$$

Depending upon the number of the terms used in the Gregory formula we successively obtain the *rectangular* (first term only), *trapezoidal* (first two terms), *parabolic* (first three terms) and *higher approximations* of the convolution integral.

Rectangular approximation. Applying [2.33], the output value for $t = n\Delta t$ becomes:

$$g_n = g(n\Delta t) = \Delta t \sum_{i=0}^{n} x_i h_{n-i} + e_{1n}$$

where all terms from the right-hand side of [2.33], except the first one, are included in e_{1n} which is the nth term of an *error sequence*. The rectangular approximation of g_n is:

$$g_{1n} = g_n - e_{1n} = \Delta t \sum_{i=0}^{n} x_i h_{n-i} \quad [2.34]$$

According to Vích (1968), the error terms e_{1n} are proportional to the sampling interval Δt. In terms of the z-transform [2.34] becomes:

$$G_1(z) = \Delta t \, X(z) \, H(z)$$

where $X(z) = \mathcal{Z}(x_n)$ and $H(z) = \mathcal{Z}\{h_n\}$. The corresponding system function is:

APPROXIMATION OF ANALOG SYSTEMS

$$\tilde{H}_1(z) = \frac{G_1(z)}{\Delta t\, X(z)} = H(z)$$

Comparing this relation with that presented in the preceding section, we can see that the impulse-invariance and rectangular-approximation output-invariance techniques provide the same results.

Trapezoidal approximation. Considering the first two terms in the integration formula we have the recursion relation:

$$g_{2n} = g_n - e_{2k} = \Delta t\, [\sum_{i=0}^{n} x_i h_{n-i} - (x_o h_n + x_n h_o)/2] \qquad [2.35]$$

The error samples e_{2n} are proportional to $(\Delta t)^2$. Applying the z-transform to both sides of [2.35] we get:

$$G_2(z) = \Delta t\, [\, X(z)\, H(z) - x_0\, H(z)/2 - X(z)\, h_0/2\,]$$

Since we consider input signals which start from zero at the time origin, $x(t) = 0$ for $t = 0$, then also $x_0 = 0$ and the z-transform $G_2(z)$ may be simplified as:

$$G_2(z) = \Delta t\, X(z)\, [H(z) - h_0/2]$$

Evidently, the related system function becomes:

$$\tilde{H}_2(z) = \frac{G_2(z)}{\Delta t\, X(z)} = H(z) - h_0/2$$

Similarly, formulae corresponding to higher approximations of the convolution integral may be derived. Comparing the expressions for individual system functions we can see that $\tilde{H}_i(z)$ may be significantly simplified and defined solely in terms of $H(z)$ and $h_0, h_1, \ldots h_{i-2}$, provided that the first input samples $x_0, x_1, \ldots, x_{i-2}$ are zeros. It is interesting that when h_0 is also zero, then $H(z) = \tilde{H}_1(z) = \tilde{H}_2(z)$. In this case $(x_0, h_0 = 0)$ the rectangular and trapezoidal approximations provide the same results. In Table IV we summarize the recursion relations, system functions and error-term estimates for approximations containing first and second differences. For some special requirements higher approximations might be necessary but it should be emphasized that for a broad range of geophysical problems, the simple rectangular approximation is usually sufficiently accurate.

2.5.3 Approximation by means of bilinear transformation

Due to the periodicity of the system function, values of $H(z)$ evaluated along the unit circle in the z^{-1}-plane will in general differ from those of $H(p)$ evaluated along the $j\omega$-axis in the p-plane, especially for frequencies close to ω_N. Let us repeat here that the periodicity property originated in the substitution $z = e^{p\Delta t}$. This mapping function transforms any complementary strip, of the width ω_s, in the

TABLE IV

Four different convolution approximations based upon the Gregory integration formula (after Vích, 1968)

Approximation	Recursion relation	System function	Error terms
Rectangular	$\dfrac{g_{1n}}{\Delta t} = \sum\limits_{i=0}^{n} x_i h_{n-i}$	$\tilde{H}_1(z) = H(z)$	$e_{1n} \sim \Delta t$
Trapezoidal	$\dfrac{g_{2n}}{\Delta t} = \sum\limits_{i=0}^{n} x_i h_{n-i} - \dfrac{1}{2}(x_0 h_n + x_n h_0)$	$\tilde{H}_2(z) = H(z) - \dfrac{1}{2} h_0 \text{ for } x_0 = 0$	$e_{2n} \sim (\Delta t)^2$
Parabolic (first differences applied)	$\dfrac{g_{3n}}{\Delta t} = \sum\limits_{i=0}^{n} x_i h_{n-i} - \dfrac{7}{12}(x_0 h_n + x_n h_0)$ $+ \dfrac{1}{12}(x_1 h_n + x_{n-1} h_1)$	$\tilde{H}_3(z) = H(z) - \dfrac{7}{12} h_0 + \dfrac{1}{12} h_1 z^{-1}$ for $x_0, x_1 = 0$	$e_{3n} \sim (\Delta t)^3$
Second differences applied	$\dfrac{g_{4n}}{\Delta t} = \sum\limits_{i=0}^{n} x_i h_{n-i} - \dfrac{5}{8}(x_0 h_n + x_n h_0)$ $+ \dfrac{1}{6}(x_1 h_{n-1} + x_{n-1} h_1)$ $- \dfrac{1}{24}(x_2 h_{n-2} + x_{n-2} h_2)$	$\tilde{H}_4(z) = H(z) - \dfrac{5}{8} h_0 + \dfrac{1}{6} h_1 z^{-1}$ $- \dfrac{1}{24} h_2 z^{-2}$ for $x_0, x_1, x_2 = 0$	$e_{4n} \sim (\Delta t)^4$

APPROXIMATION OF ANALOG SYSTEMS

p-plane into one of the sheets of the Riemann surface in the z^{-1}-plane. To circumvent the aliasing problem we shall introduce a new transform which will map the entire left half of the p-plane (not only one strip of the width of ω_s) into the exterior of the unit circle in the z^{-1}-plane. Consequently, frequencies will be limited to the range from $-\omega_N = -\pi/\Delta t$ to $\omega_N = \pi/\Delta t$ but on the other hand the differences ascribed to the periodic property will disappear since no periodicity takes place. The process of required mapping can be realized by means of the *bilinear transformation:*

$$p = \frac{1 - z^{-1}}{1 + z^{-1}} = \frac{z - 1}{z + 1} \qquad [2.36]$$

which yields the system function of the digital filter in a form:

$$H(z) = [H(p)]_{p = \frac{1 - z^{-1}}{1 + z^{-1}}}$$

For more information on the transformation the reader is referred to Tou (1959, pp. 465–479), Rader and Gold (1967), Båth (1968, pp. 36–38) among others.

An important property of the bilinear transformation is that it preserves $H(z)$ in a form of two polynomials in z^{-1} (or in z), provided that $H(p)$ is also a ratio of two polynomials in p, which is usually the case. Both the numerator and denominator of $H(z)$ result to be of the same degree. The inverse transformations:

$$z^{-1} = \frac{1 - p}{1 + p} \quad \text{and} \quad z = \frac{1 + p}{1 - p} \qquad [2.37]$$

describe the process of mapping from the p-plane into the z^{-1} and z-planes, respectively. The imaginary axis $j\omega$ is replaced by the unit circle in both the z^{-1}- and z-planes. The mapping of several important points is shown in Table V. An interesting and important feature of the bilinear transformation follows from the table. Points distributed along the imaginary axis in the p-plane are mapped in the

TABLE V

Mapping from the p-plane into z^{-1}- and z-planes by means of the bilinear transformation

p-plane	z^{-1}-plane	z-plane
0	1	1
−1	∞	0
1	0	∞
j	$-j$	j
$-j$	j	$-j$
∞	−1	−1
imaginary axis	unit circle	unit circle
real axis	real axis	real axis

z^{-1}-plane onto points on the unit circle. Let us move a point along the unit circle from a position $C_1 = 1$ to $C_2 = -j$. As mentioned previously, these two points represent the 0- and $\omega_N/2$-value of the digital-frequency variable which we here denote by $\omega_D \Delta t$. The corresponding point in the p-plane will follow the imaginary axis between points 0 and j thus representing successively all values of the analog-frequency variable ω_A in the range $0 - 1$. Further movement of the point along the unit circle from $C_2 = -j$ to $C_3 = -1$, i.e. from $\omega_D \Delta t = \omega_N/2$ to $\omega_D \Delta t = \omega_N$, shifts the point in the p-plane from j to $j\infty$. The evidently nonlinear relation between ω_A and $\omega_D \Delta t$ is, unfortunately, the price to be paid for avoiding the folding property.

For points on the imaginary axis we have:

$$p = j\omega_A = \frac{1 - \exp(-j\omega_D \Delta t)}{1 + \exp(-j\omega_D \Delta t)} = j \tan(\omega_D \Delta t/2)$$

or: $\omega_A = \tan(\omega_D \Delta t/2)$ $\omega_D = (2/\Delta t) \arctan \omega_A$ [2.38]

Thus, when the frequencies ω_A and ω_D are related according to [2.38], functions $H\{p = j\omega_A\}$ and $H\{z = \exp(j\omega_D \Delta t)\}$ will have the same value. Some authors call ω_A and ω_D the *fictitious (pseudo) frequency* and *actual frequency*, respectively. It is also possible to compensate the nonlinear-frequency scaling by prewarping the frequency scale in the analog domain. This kind of compensation is especially convenient for filters with amplitude characteristics containing only distinct pass and rejection bands with well-defined cut-off frequencies. These frequencies may be modified according to [2.38] and the digital filter $H(z)$ is then received from $H(p)$ by replacing p by $(1 - z^{-1})/(1 + z^{-1})$. A direct application of this substitution, i.e. without prewarping, would shift the filtering properties of $H(z)$ to incorrect frequencies.

Application of the bilinear transformation in the design of digital systems which sufficiently resemble known analog systems may be summarized in the following two points:

(1) Transform the critical (cut-off, resonance, etc.) frequencies ω_{iD}, where $i = 1, 2, \ldots$, of the given analog transfer function, $H(p)$, into the fictitious-frequency domain by using the relation:

$\omega_{iA} = \tan(\omega_{iD} \Delta t/2)$

(2) To find the required digital system $H(z)$, replace p in $H(p)$ by $(1 - z^{-1})/(1 + z^{-1})$.

Blackman (1965, pp. 74–75) calls this method *frequency-transformation method*.

The whole procedure may be well illustrated by means of a numerical example. For the sake of simplicity, consider an analog system:

$$H(p) = \frac{1}{p + \Omega}$$

The amplitude response of the system is approximately flat in the frequency range $0 < \omega \leq \Omega$ in rad/sec and decreases with a slope of -20 dB/decade for frequencies $\omega > \Omega$. The true response does not deviate from this approximation by more than 3 dB, with the deviation of 3 dB observed at the corner frequency $\omega = \Omega$. Instead of the amplitude response $|H(p)|$ rather the $20 \log |H(p)|$ will be used. Further, when the logarithmic scale is used for ω, we obtain the so-called *log-modulus plot* which is commonly used by control engineers. These plots are easy to construct since they usually may be approximated, with sufficient accuracy, by a series of straight-line segments.

The procedure for deriving the system function will be the following. Firstly, we transform the corner frequency according to [2.38] into the fictitious-frequency domain:

$$\Omega \rightarrow \tan(\Omega \Delta t/2) = \omega_A$$

Secondly, we apply the substitution [2.36] which yields the required digital system:

$$H(z) = \frac{1}{(1 - z^{-1})/(1 + z^{-1}) + \tan(\Omega \Delta t/2)}$$

which to some extent will simulate the behaviour of the given analog system $H(p)$. Suppose we want to suppress all frequencies above 0.5 cps and we choose the sampling period $\Delta t = 0.1$ sec. Then:

$$\Omega = 3.14 \text{ rad/sec} \qquad \omega_A = \tan(\Omega \Delta t/2) = 0.16 \text{ rad/sec}$$

The corresponding system function becomes:

$$H(z) = \frac{1}{(1-z^{-1})/(1+z^{-1}) + 0.16} = \frac{1 + z^{-1}}{0.84(1.38 - z^{-1})}$$

The function $H(z)$ has a single zero at $\alpha_1 = -1$ and a single pole at $\beta_1 = 1.38$ and so the corresponding amplitude characteristics may easily be constructed by utilizing the pole-zero technique.

Kaiser (1966) presents the application of the bilinear transformation in the design of a special category of low-pass filters. The bilinear transformation has been employed by Neunhöfer (1971) in determining the seismograph transfer function. Other examples may be found elsewhere.

2.6 PHASE-DISTORTIONLESS FILTERS

The phase characteristics of the recursive filter may be expressed as:

$$\phi(\omega) = \arctan \frac{\text{Im } H(e^{j\omega \Delta t})}{\text{Re } H(e^{j\omega \Delta t})} \qquad [2.39]$$

where according to [2.7]:

$$H(z) = H(e^{j\omega \Delta t}) = \frac{\sum\limits_{n=0}^{M} a_n e^{j\omega n \Delta t}}{1 + \sum\limits_{n=1}^{L} b_n e^{j\omega n \Delta t}}$$

As follows from [2.39] a recursive filter in general does not have zero or linear phase characteristics which means that the filter provides a phase-distorted output. In general, signal distortions due to amplitude characteristics are necessary and commonly desirable while additional distortions due to phase characteristics are undesired and should be eliminated. It is one of the advantages of digital filters that they may rather easily be constructed with zero-phase characteristics. Below we shall discuss several design possibilities.

Consider a digital system defined by the system function:

$$H_1(z) = \frac{\sum\limits_{n=0}^{M} a_n z^{-n}}{1 + \sum\limits_{n=1}^{L} b_n z^{-n}}$$

where M, L are finite integers. As follows, e.g. from [1.39], the complex function $H_1(z) = H_1(e^{j\omega \Delta t})$ can be viewed in terms of vector summation for any angular frequency ω. For any particular ω, the magnitude and direction of the resulting vector yield the amplitude and phase responses, respectively. In order to eliminate the phase response completely, we have to introduce an additional operation which would keep the resulting vector permanently, for any ω, in the direction of the real axis. In this connection use can be made of a system function similar to $H_1(z)$, namely:

$$H_2(z) = \frac{\sum\limits_{n=0}^{M} a_n z^n}{1 + \sum\limits_{n=1}^{L} b_n z^n}$$

When graphically constructing $H_1(z)$, individual vector components turn in the clockwise direction while in the case of $H_2(z)$ these turn in the counter-clockwise direction. Consequently, $H_1(z)$ and $H_2(z)$ have the same amplitude characteristics and their phase characteristics differ in sign only. The usefulness of $H_2(z)$ is obvious when the product of both functions is considered. The product $H_1(z) H_2(z)$ yields a system function with zero phase and amplitude characteristics $|H_1(z)|^2$. In other words, cascade arrangement of filters $H(z)$ and $H(z^{-1})$ provides phase-distortionless transmission and frequency selectivity which is a square of that of the basic filter. In cases when the shape of the amplitude response is also important, $H_1(z)$

PHASE-DISTORTIONLESS FILTERS

and $H_2(z)$ have to be chosen so that $|H_1(z)| = |H_2(z)| = \sqrt{|H(z)|}$, where $H(z)$ is the desired filter. Note that if $H_1(z)$ is stable, $H_2(z)$ will be unstable and vice versa (see also Fig. 1.7). Hence, this technique may be successfully applied to finite-length signals only.

The serial arrangement of the two filters may as well be investigated analytically. The system function of the resulting filter becomes:

$$H(z) = H_1(z) H_2(z) = \frac{\sum\limits_{n=0}^{M} a_n z^{-n}}{1 + \sum\limits_{n=1}^{L} b_n z^{-n}} \times \frac{\sum\limits_{n=0}^{M} a_n z^{n}}{1 + \sum\limits_{n=1}^{L} b_n z^{n}}$$

Performing the multiplication we find that:

$$H(z) = \frac{a_M z^M + (a_M a_1 + a_{M-1}) z^{M-1} + \ldots +}{1 + b_L z^L + (b_L b_1 + b_{L-1}) z^{L-1} + \ldots +}$$

$$\frac{+ \sum\limits_{n=0}^{M} a_n^2 + \ldots + (a_{M-1} + a_M a_1) z^{-M+1} + a_M z^{-M}}{+ \sum\limits_{n=1}^{L} b_n^2 + \ldots + (b_{L-1} + b_L b_1) z^{-L+1} + b_L z^{-L}}$$

Since exponents of z which differ only in sign correspond to the same coefficients, both the numerator and denominator of $H(z)$ form symmetric power series where $a_n = a_{-n}$ for $n = 1, 2, \ldots, M$ and $b_n = b_{-n}$ for $n = 1, 2, \ldots, L$. After the substitution $z = e^{j\omega \Delta t}$ the numerator as well as the denominator become real functions of angular frequency. Consequently, $H(z)$ becomes also a real function, i.e. its phase is zero for all frequencies.

It is interesting to investigate the behaviour of $H_1(z)$ and $H_2(z)$ in more detail. For this purpose consider a simple system:

$$H_1(z) = \frac{a_0 + a_1 z^{-1}}{1 + b_1 z^{-1} + b_2 z^{-2}}$$

and a short input sequence $x = (x_0, x_1)$. Members of the output sequence, $y = (y_0, y_1, y_2)$ are determined by comparing the like powers of z in the equation:

$$Y(z) = (a_0 + a_1 z^{-1}) X(z) - (b_1 z^{-1} + b_2 z^{-2}) Y(z) \tag{2.40}$$

where $X(z) = \mathcal{Z}\{x_n\}$ and $Y(z) = \mathcal{Z}\{y_n\}$. Consider now the time-reverse input $\bar{x} = (x_1, x_0)$. The corresponding output sequence is obtained from [2.40] after the substitution $X(z) = x_1 + x_0 z^{-1}$. In terms of x_0 and x_1 the first three output samples are:

$y_0 = a_0 x_1$
$y_1 = a_0 x_0 + a_1 x_1 - b_1 y_0 = a_0 x_0 + a_1 x_1 - a_0 b_1 x_1$
$y_2 = a_1 x_0 - b_1 y_1 - b_2 y_0 = a_1 x_0 - a_0 b_1 x_0 - a_0 b_1 x_1 + a_0 b_1^2 x_1 - a_0 b_2 x_1$

As a next step we shall investigate the transmission of the system $H_2(z)$. Due to the positive exponents of z, [2.40] will be slightly modified so that the input–output relation in the z domain becomes:

$$U(z) = (a_0 + a_1 z) X(z) - (b_1 z + b_2 z^2) U(z) \qquad [2.41]$$

where $U(z) = \mathcal{Z}\{u_n\}$.

Multiplication by z physically means time advance by one sampling interval. Accordingly, the system $H_2(z)$ contains advance blocks in both the forward and feedback paths. Output terms will therefore depend upon the present and future input and also upon the future output terms. Dependence upon the future input makes the system physically nonrealizable, an obstacle which may easily be overcome when data in a stored form are to be processed. The dependence upon the future output values creates an additional difficulty since the present output is also influenced by terms which have not yet been determined. Nevertheless, a simple trick makes the procedure computationally realizable. Instead of calculating the output sequence u in the direction of increasing time, we apply the reverse approach starting with the output term which corresponds to the last term of the input sequence. From [2.41] we obtain the recursion relation:

$$u_n = a_0 x_n + a_1 x_{n+1} - b_1 u_{n+1} - b_2 u_{n+2}$$

Since we applied the input signal $x = (x_0, x_1)$, we determine the output terms in the sequence $u_1, u_0, u_{-1}, \ldots$. In terms of the input values the last three output values are:

$$u_1 = a_0 x_1$$
$$u_0 = a_0 x_0 + a_1 x_1 - b_1 u_1 = a_0 x_0 + a_1 x_1 - a_0 b_1 x_1$$
$$u_{-1} = a_1 x_0 - b_1 u_0 - b_2 u_1 = a_1 x_0 - a_0 b_1 x_0 - a_1 b_1 x_1 + a_0 b_1^2 x_1 - a_0 b_2 x_1$$

Comparison of output sequences of filters $H_1(z)$ and $H_2(z)$ reveals an interesting and important phenomenon. Replacing the system $H_1(z)$ by $H_2(z)$ is equivalent to passing the time-reversed input through $H_1(z)$ and reversing the received output again. In other words, the two output sequences y and u are identical when one of them is time-reversed prior to the term-by-term comparison. For example, the first term of y is equal to the last term of u, etc. Due to this property the system $H_2(z)$ is sometimes called the *reverse-time filter*.

Another form of phase-distortionless filters may be realized via summation or subtraction of output sequences of $H_1(z)$ and $H_2(z)$. In order to explain the principles of the method, consider simple systems as:

$$H_1(z) = 1 - z^{-1} \qquad\qquad H_2(z) = 1 - z$$

and an arbitrary input sequence $x = (x_0, x_1, x_2, \ldots)$. The z-transforms of the output signals are:

$$Y_1(z) = (1 - z^{-1}) X(z) \qquad Y_2(z) = (1 - z) X(z)$$

where $X(z) = \mathcal{Z}\{x_n\}$. The corresponding recursion relations become:

$$y_{1n} = x_n - x_{n-1} \qquad y_{2n} = x_n - x_{n+1}$$

Point-by-point summations and subtractions of these two outputs yield:

$$u_{1n} = y_{1n} + y_{2n} = -x_{n+1} + 2x_n - x_{n-1}$$
$$u_{2n} = y_{1n} - y_{2n} = x_{n+1} - x_{n-1}$$

It is evident that the resulting sequences u_1 and u_2 are outputs of filters:

$$U_1(z) = z + 2 - z^{-1} \qquad U_2(z) = z - z^{-1}$$

respectively, provided that they are excited with the same input signal $x = (x_0, x_1, x_2, \ldots)$. It follows directly from their form that both $U_1(z)$ and $U_2(z)$ are symmetric filters (even and odd, respectively), hence producing phase-distortionless transmission.

Let us now investigate the relationship between the original system $H_1(z)$ and systems $U_1(z)$ and $U_2(z)$. To do this we substitute $z = e^{j\omega\Delta t}$ back into the expression for $H_1(z)$ which gives:

$$H_1(z) = 1 - z^{-1} = 1 - e^{-j\omega\Delta t} = 1 - \cos \omega\Delta t + j \sin \omega\Delta t$$

Separating the real and imaginary parts we have:

$$2 \operatorname{Re} H_1(z) = 2(1 - \cos \omega\Delta t) = 2 - e^{j\omega\Delta t} - e^{-j\omega\Delta t} = z + 2 - z^{-1}$$
$$2 \operatorname{Im} H_1(z) = 2 \sin \omega\Delta t = -j(e^{j\omega\Delta t} - e^{-j\omega\Delta t}) = -j(z - z^{-1})$$

Simple comparison reveals that:

$$U_1(z) = 2 \operatorname{Re} H_1(z) \qquad U_2(z) = j 2 \operatorname{Im} H_1(z)$$

The results, which have general validity, may be summarized in the following points (see also Mooney, 1968). Firstly, a parallel point-by-point summation $y_{1n} + y_{2n}$, substitutes a zero-phase filter by a system function $2 \operatorname{Re} H_1(z)$. Secondly, a parallel subtraction, $y_{1n} - y_{2n}$, substitutes a 90°-phase filter by a system function $j 2 \operatorname{Im} H_1(z)$.

There is no doubt that filters $U_1(z)$ and $U_2(z)$ perform a perfect phase-distortionless transmission. Nevertheless, in each particular case one has to check to what extent $\operatorname{Re} H_1(z)$ and $\operatorname{Im} H_1(z)$ preserve the desired filtering properties. Whereas in the serial arrangement the amplitude characteristics, $|H_1(z)|^2$, are available in a convenient form, real and imaginary parts of $H_1(z)$ usually do not offer a direct interpretation in terms of selective properties of the system. The former approach is suitable for short, the latter for long input sequences.

There is a large category of filters with only complex zeros all placed on the unit circle (see e.g. the simple $0.25\omega_N$-rejection filter discussed in Section 2.4). Such a distribution of zeros of $H(z)$ is to some extent advantageous when a phase-distortionless filter is to be designed. Consider a system function in its general form:

$$H(z) = \frac{A(z)}{B(z)} = \frac{a_0 + a_1 z^{-1} + \ldots + a_M z^{-M}}{1 + b_1 z^{-1} + \ldots + b_L z^{-L}}$$

As has been shown in connection with Fig. 2.6, the poles of rejection filters are usually located close to zeros in order to preserve the amplitude characteristics approximately constant for frequencies outside the rejection band. Thus, the polynomials in the numerator and denominator of $H(z)$ are usually of the same degree, $M = L$. The input–output relation yields:

$$Y(z) = \frac{A(z)}{B(z)} X(z) \qquad [2.42]$$

where $Y(z)$ and $X(z)$ are the z-transforms of output and input signals, respectively.

Rewriting [2.42] in the same way as has been done in [2.19] we have:

$$Y(z) = U(z) A(z)$$

where $U(z) = X(z)/B(z)$. This separation may be interpreted as a substitution of a serial arrangement of a convolution filter, $A(z)$, and a recursive filter, $1/B(z)$, for the recursive filter given by [2.42]. Since all the zeros of $H(z)$ are assumed to be located on the unit circle, the numerator of the system function may be expressed as:

$$A(z) = \prod_{i=1}^{M/2} a_0 \, [z^{-1} - \exp(-j\Omega_i)] \, [z^{-1} - \exp(j\Omega_i)] \qquad [2.43]$$

where $\exp(\pm j\Omega_i)$ is the ith pair of the $M/2$ pairs of complex-conjugate zeros of $H(z)$. In order to prevent the existence of imaginary output values, complex zeros, similarly to complex poles, appear only in complex-conjugate pairs. Hence, in our case, M is evidently an even number. If we consider separately the ith pair in [2.43], we have a product:

$$[z^{-1} - \exp(-j\Omega_i)] \, [z^{-1} - \exp(j\Omega_i)] = z^{-2} - 2 \operatorname{Re}[\exp(j\Omega_i)] z^{-1} + 1$$

which is always a symmetric second-degree polynomial in z^{-1}. The numerator $A(z)$, being a product of such polynomials, is again symmetric.

Consequently, the convolution with $A(z)$ is free of any phase distortion. Due to the stability requirements the poles must not be located on the unit circle. The denominator, $B(z)$, forms an asymmetric polynomial, hence introducing a non-linear phase shift. The phase distortion due to the recursive part, $1/B(z)$, of the filter $H(z)$ may be eliminated by making use of the time-inverted signals as discussed above. The input signal passes first through the recursive filter $1/B(z)$, then through the corresponding reverse-time filter, and finally through the convolution filter $A(z)$. Note that the resulting filter differs from that defined in [2.42]. While the originally required amplitude characteristics are:

$$|H(z)| = |A(z)/B(z)|$$

the phase-distortionless modification provides:

$$|H_1(z)| = |A(z)|/|B(z)|^2$$

and accordingly, the output sequence $\{y_{1k}\}$ is produced instead of $\{y_k\}$. Special care has to be taken to assure that the realized system $H_1(z)$ satisfies the required filtering properties of $H(z)$. The procedure seems to be rather complicated when compared with preceding methods and is therefore more of theoretical than practical use (see also Göncz and Zelei, 1972).

2.7 EFFECTS OF QUANTIZATION IN DIGITAL FILTERS

According to definition, analog quantities may take on a continuous range of amplitude values, measurable with an arbitrary number of significant digits. In practice, this is limited by the accuracy of the measuring device used but the measured quantity itself does not create any restriction. On the other hand, the term digital always implies quantized values, i.e. the amplitudes of quantities are available in a set of levels differing by finite steps, say E, known as the *quantization interval*. In other words, the true amplitude is approximated by the closest quantizing value. Requirements governing the choice of the magnitude of E have been discussed by Zürn (1974). Since digital computers process only a finite number of significant digits, there is always only a finite set of levels for any particular number or amplitude to be processed by the computer. This means that digital-filter parameters as well as the filtering process, realized by a computational algorithm, are available with finite accuracy only (see e.g. Donnell, 1967; Bendat and Piersol, 1971). The quantized and exact values differ from each other by an amount which does not exceed $E/2$. The finite-word-length constraint inherent to any digital computer introduces errors of three essential types. Firstly, errors due to the quantization of the input signal, secondly, errors caused by rounding off the results of arithmetic operations and thirdly, errors following from the quantization of filter parameters. These errors are treated separately below.

2.7.1 *Errors caused by input quantization*

When the input signal is of a digital nature from the beginning, then evidently no errors of this type exist. By contrast, when the input signal $x(t)$ is of analog form, the analog–digital conversion has to be carried out prior to computer processing. The output of the converter is the quantized signal x_n with a form of a staircase wave.

Thus, the process of this conversion is a basic source of errors of the first type. For example, for analog amplitudes equal to 0, $\pm E$, $\pm 2E$, ..., any change by less than $\pm E/2$ results in the same quantized value. When quantization errors are small, compared with the amplitudes of $x(t)$, the error introduced by the analog–digital conversion has about the same effect as a noise superimposed on $x(t)$. In this respect, the quantized input is written in a form:

$$x_n = x(n\Delta t) + q(n\Delta t)$$

where $x(n\Delta t)$ is the noiseless analog input, sampled with the sampling period Δt, $q(n\Delta t)$ is the added noise and n is an integer. Assuming that samples $q(n\Delta t)$ are mutually independent, and have a uniform probability distribution, $p(q) = 1/E$, within the interval $-E/2 \leq q(n\Delta t) \leq E/2$ and zero distribution otherwise, the variance of the error is:

$$\overline{q^2} = \int_{-\infty}^{\infty} q^2 p(q)\, dq = \frac{1}{E} \int_{-\frac{E}{2}}^{\frac{E}{2}} q^2\, dq = \frac{E^2}{12}$$

In order to test the influence of input-quantization errors upon any particular system, one of the two following approaches may be utilized. Provided that the digitization of the input signal is sufficiently fine, $q(n\Delta t)$ is treated as an additive random noise. Quantization of the input changes its waveform slightly, but otherwise it usually does not create new processing problems. For simple systems it is suitable to calculate analytically the characteristics of the output signal resulting from the input noise $q(n\Delta t)$. Otherwise, it is possible to simulate the noise transmission on a computer by making use of a random-noise generator. Kulhánek and Klíma (1970) investigated errors of the first type in the frequency domain. They used analytical Berlage pulses and estimated frequency ranges for reliable analog–digital conversion, assuming a particular digitizing device and spectral shape of the input signal.

2.7.2 Errors caused by product quantization

A second source of errors, the quantization or rounding off the products resulting during the signal processing, represents a more severe and complicated obstacle. Detailed discussion of the subject comprising the direct- and canonic-form realization together with numerical examples leading to specification of word length are presented by Rader and Gold (1967). Below we describe the principles of investigation of some of the statistical characteristics of this type of error.

To illustrate the problem we consider a simple first-order recursive filter. For this purpose suppose a recursion formula:

$$y_n = x_n - 0.5 y_{n-1}$$

and the input sequence:

$x = (1, 1, 1, \ldots)$

As a result of one multiplication and one subtraction (see the recursion formula), first several coefficients of the output sequence are:

$y = (1, 0.5, 0.75, 0.625, 0.6875, 0.65625, 0.671875, \ldots)$

The numerical example makes it evident that every iteration step increases the number of decimal places of the exact value of y_n by one. Further, it is clear that the increase of decimal places per iteration step is ruled by the given accuracy (number of decimal places) of the coefficients in the feedback path of the filter. Since in our case $b_1 = 0.5$ we observe a rate of increase of one decimal place per iteration step. When the number of output terms increases without limit, $n \to \infty$, an exact representation of the output sequence $\{y_k\}$ of even such a simple filter as mentioned above would require processing of infinite-length numbers. Due to the fact that no digital computer can fulfill such a requirement, the errors ascribed to rounding off the results of multiplication in digital filters are inevitable.

It is worth noting, and it follows also from the previous example, that this type of error is due to the feedback action inherent to any type of recursive filter. Consequently, nonrecursive filters which contain only the direct path are in general not affected by the quantization of results of multiplication. It is the part of the output signal, recirculated back into the input, which gives rise to round-off errors. Hence, the true effect on the output signal of rounding off the products appearing in the iteration process depends also on the filter realization. Consider a general direct-form representation for the Lth-order recursive filter, $H(z) = A(z)/B(z)$, diagrammed in Fig. 2.3a and let the quantization-error sequence be r_n. Then, according to results presented in Section 2.2.3 (see also Rader and Gold, 1967), the recursion relation for the output sequence affected by the quantization noise becomes:

$$y_n = \sum_{i=0}^{M} a_i x_{n-i} - \sum_{i=1}^{L} b_i y_{n-i} + r_n$$

where M and L are degrees of polynomials in the numerator and denominator of $H(z)$, respectively. The z-transform of both sides of this equation yields:

$$Y(z) = X(z) \frac{\sum_{i=0}^{M} a_i z^{-i}}{1 + \sum_{i=1}^{L} b_i z^{-i}} + R(z) \frac{1}{1 + \sum_{i=1}^{L} b_i z} \qquad [2.44]$$

Equation [2.44] describes the z-transform of the output as a sum of two terms. From the comparison with formulae presented in Section 2.2.3 it is evident that the first term corresponds to the transmission of the sequence $\{x_n\}$ through $H(z)$ which is free of quantization errors. The second term describes the transmission of the error sequence $\{r_n\}$ through $1/B(z)$ which is the denominator (feedback part) of $H(z)$. As in the preceding case, the quantization of multiplication results has a similar effect as a noise superimposed on the noiseless output signal. Generally speaking, round-off errors could appear at the output of blocks in the direct as well as in the feedback part of the network (Fig. 2.3a). In the former case, it is the number of significant bits of the input samples and coefficients $a_0, a_1, ..., a_M$ relative to the word length of the computer used which decides whether round-off takes place or not. Hereafter, this type of errors will be neglected. In the latter case, however, the rounding-off is inescapable due to the inherent feedback character of any recursive filter. Errors appearing at the output of individual feedback blocks may be replaced by a single noise source, $\{r_n\}$, injected at the filter output, $\{y_n\}$, as has been shown in [2.44]. For the filter realization considered, the mean-square value of the noise component of the output sequence $\{y_n\}$ may be estimated in terms of the quantization step size, E. For this purpose we shall here consider only the second term on the right-hand side of [2.44]. It follows from this equation that the output noise may be visualized as the output of the system $1/B(z)$ which has been excited by the input sequence $\{r_n\}$.

In order to be able to estimate the behaviour (in a statistical sense) of the output noise we shall need some of the statistical characteristics, namely the mean-square value expressed in terms of the autocorrelation or power spectrum. For the sake of simplicity let us assume the successive samples of the input noise be mutually uncorrelated white noise. For such random signals the autocorrelation function at the delay number m becomes:

$$R_{rr}(m) = \begin{cases} \lim_{N \to \infty} \frac{1}{2N+1} \sum_{n=-N}^{N} r_n r_{n+m} = \frac{E^2}{12} = \overline{r^2} & \text{for } m = 0 \\ 0 & \text{otherwise} \end{cases} \quad [2.45]$$

where $2N + 1$ is the number of terms of the noise sequence $\{r_n\}$. Realizing that the autocorrelation function, $R_{rr}(m)$, and the corresponding power spectrum, $\Phi_{rr}(z)$, form a transform pair, so that

$$\Phi_{rr}(z) = \sum_{m=-\infty}^{\infty} R_{rr}(m) z^{-m} = \frac{E^2}{12}$$

an alternative expression for the mean-square value $\overline{r^2}$ becomes:

$$\overline{r^2} = R_{rr}(0) = \frac{1}{2\pi j} \oint_\Gamma \Phi_{rr}(z) z^{-1} \, dz \quad [2.46]$$

The power spectrum, $\Phi_{rr}(z)$, is constant in the entire frequency range, because of our assumption of white noise. If the noise-power spectrum at the input of the system $1/B(z)$ is equal to $E^2/12$ then, according to [1.23], the output-power spectrum can be written as:

$$\Phi_{nn}(z) = \frac{1}{B(z) B(z^{-1})} \frac{E^2}{12}$$

and [2.46] yields the mean-square value of the output noise:

$$\overline{n_r^2} = \frac{E^2}{24\pi j} \oint_\Gamma \frac{z^{-1} dz}{B(z) B(z^{-1})} \qquad [2.47]$$

Obviously, the value of $\overline{n_r^2}$ is determined by the quantization-step size, E, and by the location of the poles of the system function $H(z)$.

A similar procedure may be applied when a canonic-form realization is to be utilized. It follows immediately from Fig. 2.3b that for this type of block-diagram representation, it is the sequence $\{u_n\}$ which is recirculated back into the input and which is used in further computation rather than the sequence $\{y_n\}$ as was the previous case. Therefore, the effect of quantization may be visualized by means of a noise signal added to the signal u. As in the case of the direct realization, the individual noise sources in the feedback part of the network have here been replaced by a single noise source. Using [2.19] and [2.20] we write the transformed input–output relations as:

$$Y(z) = U(z) \sum_{i=0}^{M} a_i z^{-i}$$

$$U(z) = \frac{X(z) + V(z)}{1 + \sum_{i=1}^{L} b_i z^{-i}}$$

where $V(z)$ is the z-transform of the noise sequence $\{v_n\}$. Corresponding recursion relations are:

$$y_n = \sum_{i=0}^{M} a_i u_{n-i} \quad \text{and} \quad u_n = x_n - \sum_{i=1}^{L} b_i z^{-i} + v_n$$

Introducing the expression for $U(z)$ into the equation for $Y(z)$ we obtain a single relation:

$$Y(z) = \frac{[X(z) + V(z)] \sum_{i=0}^{M} a_i z^{-i}}{1 + \sum_{i=1}^{L} b_i z^{-i}} = X(z) H(z) + V(z) H(z) \qquad [2.48]$$

As in the case of the direct-form realization, [2.48] consists of two terms. The first term describes the passage of the input x through $H(z)$, which is free of quantization errors, whereas the second term determines the transmission of the noise v through the system $H(z)$. The difference between the direct- and canonic-form realization is that while in the former case the quantization noise is passing only through the feedback part (only the denominator of $H(z)$ is involved), in the latter case the noise passes through both the feedback and the direct part of the system.

In agreement with [2.47] the mean-square value of the output noise, $\overline{n_v^2}$, for the canonic-form realization is:

$$\overline{n_v^2} = \frac{E^2}{24\pi j} \oint_\Gamma H(z)\, H(z^{-1})\, z^{-1}\, dz \qquad [2.49]$$

Concluding our discussion we may state that statistical characteristics of second-type errors may be estimated by making use of auxiliary white-noise sources as additional inputs to a noiseless filter. Simple adding of the noise to the output, as has been done in the case of errors of the first type, is here no more applicable since the location of the noise source depends upon the realization of the digital system. The variance of the output-noise samples for realization in the direct and canonic form are estimated by using [2.47] and [2.49], respectively. It follows from both these equations that the value of these variances is ruled by the quantization-step size and by the location of singularities of the system function. By using a one-pole filter, Rader and Gold (1967) showed that the output variance increases rapidly when the pole position approaches the unit circle.

2.7.3. *Errors caused by filter-parameter quantization*

Errors of the third type, namely the quantization of filter parameters, initially specified with an unlimited accuracy, change the characteristics of the filter. Obviously, any deviation from the exact values of coefficients $a_0, a_1, \ldots, a_M$ and $b_1, b_2, \ldots, b_L$ will cause a change in the position of zeros and poles of the system function. The dependence upon the parameter accuracy increases with the increasing system-function degree and with poles approaching the unit circle. For example, Gold and Rader (1969) recommend limiting ourselves to first- and second-order filters only. Higher-order filters, even if designed to be stable, may easily provide unstable output signals due to the inexact values of filter parameters applied. Thus, when high-order filter is to be used, an extreme care must be taken in determining the co-efficients of the system function. An alternative, and from accuracy considerations far better, solution is to substitute the original system function by a number of first- and/or second-order subfilters. The decomposition of a ratio of two polynomials into partial fractions has been discussed in Section 2.5.1. It is recommended that the polynomial factorization be performed in the *p*-domain rather than in the *z*-domain. An interesting phenomenon is the dependence of the third-type error upon the sampling period Δt. In short, more dense sampling, i.e. smaller Δt, generally requires higher accuracy of system parameters at the same time. When quantization errors are taken into account, then, a simple decrease of Δt does not necessarily provide better approximation of the analog filter. On the contrary, it may lead to instability of the filter. Considerable details concerning this problem as well as relationships between the position of poles, filter order, sampling period and co-efficient accuracy may be found in Kaiser (1966) and Gold and Rader (1969).

Chapter 3

LOW-, HIGH- AND BAND-PASS FILTERS

The objective of this chapter is to discuss properties of three essential filter categories. The division is made according to the form of the frequency response function. Assuming *ideal response functions* the categories are the following:

(1) *Low-pass filters* whose amplitude response is constant within the pass band $|\omega| \leq \omega_L$ and zero elsewhere.

(2) *High-pass filters* whose amplitude response is constant within the pass band $|\omega| \geq \omega_H$ and zero elsewhere.

(3) *Band-pass filters* which exhibit uniform transmission within the pass band $\omega_L \leq |\omega| \leq \omega_H$ and zero transmission (infinite attenuation) outside the pass band. Frequencies ω_L, ω_H are called *cut-off frequencies*.

The response functions mentioned above and displayed in Fig. 3.1 are those of ideal filters which, for reasons described below, can only be approximated in practice. The three categories will be considered separately. Nevertheless, most of our

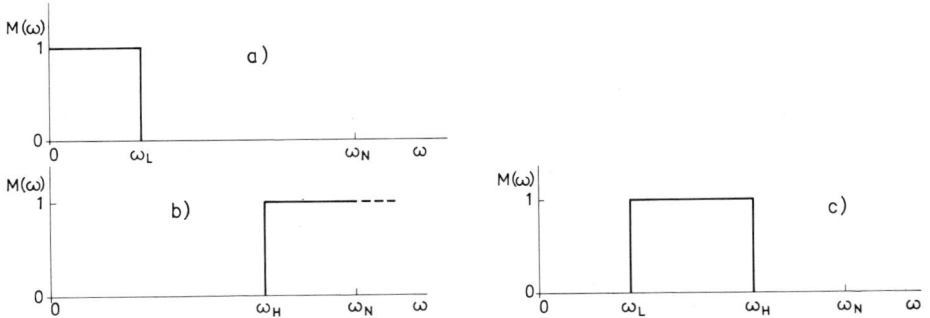

Fig. 3.1. Amplitude characteristics of basic ideal filters: (a) low-pass; (b) high-pass; (c) band-pass. Only positive frequencies are diagrammed.

attention is paid to low-pass filters since they provide the essential relations for deriving either of the other two types. To some extent, high-pass filters may be considered as an inverse of low-pass filters, and band-pass filters may be viewed as a combination of low- and high-pass filters. To complete the list, band-rejection or notch filters should be also included. However, these filters have been briefly discussed already in Chapter 2 and are therefore omitted here. Band-rejection characteristics may be determined in terms of a band-pass filter and/or as a combination of a low- and high-pass filter. Some authors (see e.g. Wood, 1968) use a common name *pass filters* for the three categories mentioned above.

3.1 LOW-PASS FILTERS

By definition, an ideal low-pass filter passes all low frequencies, $|\omega| \leq \omega_L$, without any change and blocks all high frequencies, $|\omega| > \omega_L$. To a certain extent, low-pass filtering may be associated with performing running averages as described e.g. by Swartz and Sokoloff (1954).

Negative frequencies, generally without any physical meaning, have been taken into account because of mathematical convenience. Since the system must respond by a real output to any real input the unit-impulse response must also be a real function of time and the real and imaginary part of the system function must be even and odd functions of frequency, respectively. We usually prefer filters with zero-phase shift and therefore wherever possible we shall consider the imaginary part of the system function to be zero.

3.1.1 *Ideal low-pass filters*

Consider an ideal filter:

$$H(\omega) = \begin{cases} A & \text{for } |\omega| \leq \omega_L \\ 0 & \text{for } |\omega| > \omega_L \end{cases} \qquad [3.1]$$

where A is a real constant defining the gain of the filter in the pass band. It is common to normalize the system function so that the gain becomes unity, i.e. $A = 1$. Since $H(\omega)$ is real, equal to A or zero, for any frequency ω no phase shift is involved. The unit-impulse response and $H(\omega)$ defined in [3.1] form a Fourier transform pair so that:

$$h(t) = \frac{1}{2\pi} \int_{-\omega_L}^{\omega_L} H(\omega) e^{j\omega t} d\omega = \frac{A}{\pi} \int_0^{\omega_L} \cos \omega t \, d\omega = \frac{A}{\pi t} \sin \omega_L t = \frac{A \omega_L}{\pi} \frac{\sin \omega_L t}{\omega_L t} \qquad [3.2]$$

Evidently, the function $h(t)$, defined for any t has the form of $(\sin x)/x$.

The next step will be the evaluation of terms of the weighting sequence $\{h_n\}$. Let Δt and ω_N be the sampling period and folding frequency, respectively. Assume further that the inequality $|\omega_N| > |\omega_L|$ holds. According to [1.33] the two quantities are related as:

$1/\Delta t = \omega_N/\pi$

LOW-PASS FILTERS

so that the discrete time instants are:

$n\Delta t = n\pi/\omega_N$

where n is an integer. After substituting quantized time for the continuous independent variable t and utilizing the results of [3.2], the impulse-invariant filter weights are:

$$h_n = \frac{A}{\pi} \int_0^{\omega_L} \cos(\omega n\Delta t)\, d\omega = \frac{A\omega_L}{\pi} \frac{\sin(\omega_L n\Delta t)}{\omega_L n\Delta t} \qquad [3.3]$$

where n ranges over all integers from $-\infty$ to ∞. In other words, for an exact realization of an ideal low-pass filter, described by [3.1], an infinitely long sequence $\{h_n\}$ is required. Since $\{h_n\}$ extends in the direction of both positive and negative time, the digital filter $H(z) = \mathcal{Z}\{h_n\}$ is necessarily a noncausal system.

For a phase-distortionless transmission the zero-phase response is a sufficient but not necessary condition. As has been shown in Chapter 1, in fact nothing more than a linear phase is required. Thus, an ideal phase-distortionless low-pass filter may generally be written in a form:

$$H(\omega) = \begin{cases} A\exp(-j\omega t_1) & \text{for } |\omega| \leq \omega_L \\ 0 & \text{for } |\omega| > \omega_L \end{cases} \qquad [3.4]$$

where t_1 is a time constant defining the phase response. According to the time-shifting theorem (Papoulis, 1962, pp. 14–15), the impulse response of this filter is given by:

$$h(t) = \frac{A\omega_L}{\pi} \frac{\sin \omega_L (t - t_1)}{\omega_L (t - t_1)}$$

Provided that the constant t_1 is a multiple of the sampling period, i.e. $t_1 = m\Delta t$, for the quantized time we have:

$$h_n = \frac{A\omega_L}{\pi} \frac{\sin \omega_L \Delta t (n - m)}{\omega_L \Delta t (n - m)} \qquad [3.5]$$

It follows from [3.3] that in the case of a zero-phase response, the sequence $\{h_n\}$ is symmetric about $n = 0$. In the case of linear-phase characteristics $\{h_n\}$ is again symmetric but the center of symmetry is shifted to the point $n = m$. Thus a filter described by [3.4] or [3.5] produces a certain time shift between the input and corresponding output signals. The time interval $m\Delta t$ is called *delay time* when $m > 0$ and *advance time* when $m < 0$. Zero-phase filters, $m = 0$, may be considered as a special case of the more general linear-phase filters. When $m = 0$, no time shift takes place, and the output responds immediately to any input excitation.

In order to evaluate the impulse-response sequence $\{h_n\}$, the Fourier series approach mentioned in Chapter 2 may also be utilized. For digital filters, with sampling period Δt, the system function $H(\omega)$ in [3.1] becomes a periodic function of frequency with a period $\omega_s = 2\pi/\Delta t$. According to results presented in Chapter 1 (see Fig. 1.3), the digitized $H(\omega)$ is divided into primary and complementary components and the amplitudes are reduced by a factor of $1/\Delta t$. Limiting ourselves to primary components only, an even periodic function $H_S(\omega) = H_S(\omega \pm k\omega_s)$ may be expressed as:

$$H_S(\omega) = \begin{cases} A/\Delta t & \text{for } |\omega| \leq \omega_L \\ 0 & \text{for } \omega_L < |\omega| \leq \omega_s/2 \end{cases}$$

the periodic $H_S(\omega)$ can be expressed as a sum of cosine terms:

$$H_S(\omega) = \frac{h_0}{2} + \sum_{n=1}^{\infty} h_n \cos \omega n\Delta t$$

Since components from the frequency interval $\omega_L < |\omega| \leq \omega_s/2$ do not produce any contribution, coefficients h_n become:

$$h_n = \frac{2}{\omega_s} \int_{-\omega_s/2}^{\omega_s/2} H_S(\omega) \cos \omega n\Delta t \, d\omega = \frac{2A}{\omega_N \Delta t} \int_0^{\omega_L} \cos \omega n\Delta t \, d\omega$$

$$h_n = \begin{cases} \dfrac{2A \, \omega_L}{\Delta t \, \omega_N} \dfrac{\sin \omega_L n\Delta t}{\omega_L n\Delta t} & \text{for } n = 1, 2, \ldots \\ \dfrac{2A \, \omega_L}{\Delta t \, \omega_N} & \text{for } n = 0 \end{cases} \qquad [3.6]$$

Applying the substitution $\omega_N = \pi/\Delta t$ in [3.6], the weighting coefficients h_n may be evaluated in terms of the sampling period and the cut-off frequency as:

$$h_n = \begin{cases} \dfrac{2A \, \omega_L}{\pi} \dfrac{\sin \omega_L n\Delta t}{\omega_L n\Delta t} & \text{for } n = 1, 2, \ldots \\ \dfrac{2A \, \omega_L}{\pi} & \text{for } n = 0 \end{cases} \qquad [3.7]$$

The system function $H_S(\omega)$ is a real and even function of frequency and therefore the corresponding impulse response is an even time sequence, with $h_n = h_{-n}$. As can be seen in [3.7] only nonnegative integers have been used. However, it is possible to extend the validity of [3.7] to negative integers simply by reducing the values of h_n for $n = 1, 2, \ldots$ by a factor of 0.5. Then the weighting coefficients obtained from [3.7] resemble those from [3.3] and the system function becomes:

$$H(z) = \tfrac{1}{2} \sum_{n=-\infty}^{\infty} h_n z^{-n} \qquad \text{where } h_{-n} = h_n \qquad [3.8]$$

which in fact repeats the result already obtained in [2.4].

3.1.2 *Truncated unit-impulse response function*

From the above discussion, it seems that low-pass filters with impulse responses defined by [3.3], [3.5] and [3.7] perform an ideal separation of the desired frequency components and the unwanted noise. While the signal spectrum in the frequency interval $(-\omega_L, \omega_L)$ is transmitted without any change, the noise spectrum for $|\omega| > \omega_L$ is completely eliminated. However, in practice there are several factors which make the realization of the ideal filters described in the preceding section impossible (see also Jones et al., 1955). As follows from [3.6], an ideal filter requires an infinitely long impulse response. Computer-simulated filters cannot process infinitely long sequences and it is necessary to truncate the impulse response after a certain finite number of terms. Besides other undesired effects (Toman, 1965; Jackson, 1967; Ulrych, 1972), the truncation introduces undulations, or ripples in the amplitude response in the pass band and in the stop band. Due to this ripple, the amplitude characteristics are no longer constant in the interval $|\omega| \leq \omega_L$, nor are they zero throughout the interval $|\omega| > \omega_L$ (see Båth, 1974, pp. 100–105). This means that some of the unwanted components are passed by the practical filter and some of the components in the pass band are attenuated more than others. It is known from the spectral analysis of time series (see also Section 2.1) that the ripple around the desired characteristics can be reduced by suitably weighting the coefficients h_n. A weighting function is introduced in order to smooth the abrupt truncation of $\{h_n\}$, thus suppressing the ripple of the amplitude characteristics.

When the amplitude response, $|H(\omega)|$, is continuous, the system function in [3.8] approximates rather well the behaviour of $H(\omega)$ for sufficiently large n. However, at the cut-off frequency for $\omega = \pm \omega_L$ the function $|H(\omega)|$ is discontinuous, changing its value abruptly from A to zero. As a direct consequence of the form of $|H(\omega)|$ overshoot near the discontinuities, known as the *Gibbs phenomenon*, appears in the approximate system function. The overshoot may give rise to undesired deviations from the desired response. The severity of the overshoot cannot be reduced by increasing the length of the impulse response, but it can be reduced via smoothing the sharp changes in the desired amplitude response (Meyerhoff, 1968a, b, c). Therefore, the usual procedure is first to approximate the discontinuous $|H(\omega)|$ by a continuous function which is then converted into the corresponding digital form. In most cases this implies a compromise of the initial requirement of an ideally sharp discrimination between the desired and unwanted portions of the processed signal.

Since we look for a continuous amplitude response, we have to introduce a transition-frequency interval within which the pass-band amplitude gradually decreases or rolls off to the stop-band amplitude. Fine details depend upon the particular filter design, nevertheless continuous roll-off always causes a portion of the useful signal to be attenuated and a portion of noise to be transmitted through the filter. The impulse-response function is then found from the continuous amplitude response via the inverse Fourier transform.

There are a number of methods available for constructing finite-length low-pass operators. Most of them are based upon requirements defined in the frequency domain. Some of these methods are outlined in following section of this chapter. Chan and Leong (1972, 1974) present an interesting time-domain design technique. They smooth the corrupted trace by making use of a moving least-squares polynomial of a fixed degree. Chan and Leong successfully applied the method to gravity data in order to enhance hidden anomalies. They showed that the processing by symmetric least-squares operators is equivalent to low-pass phase-distortionless filtering. The characteristics of the operator depend upon the degree of the polynomial and the number of data points. In order to increase the efficiency of the process the data are passed several times through the same filter.

Another example of a symmetric low-pass operator is *Strakhof's filter*. Naidu (1968) applied the filter to smooth aeromagnetic observational data. Smoothing by means of least-squares polynomials and corresponding z-transform behaviour has been discussed by Wood and Hockens (1970). Shapiro (1970) describes smoothing operations in terms of low-, high- and band-pass filtering for one- and two-dimensional data. Time-domain smoothing of marine gravimetric records by means of three different weighting functions has been studied by Boyarskiy and Kogan (1968).

3.1.3 *Ormsby and Martin-Graham filters*

In this subsection we briefly describe an interesting approach presented by Anders et al. (1964), which simplifies the rather laborious calculations and offers quite large freedom in choosing the shape of the roll-off.

Consider an ideal transfer function $L(\omega)$ which is an even function of frequency, unity within the pass band and zero elsewhere. The ideal impulse response, $l(t)$, is evaluated from $L(\omega)$. By specifying the roll-off we find the approximate continuous transfer function, $H(\omega)$ and the corresponding impulse response $h(t)$. If we assume that a special weighting of $l(t)$ provides $h(t)$ then we write:

$$h(t) = l(t) \, w(t) \qquad [3.9]$$

where $w(t)$ is the weighting function which is to be determined according to the required behaviour of the roll-off. Due to the convolution theorem, multiplication in the time domain corresponds to convolution in the frequency domain. Thus:

$$H(\omega) = \int_{-\infty}^{\infty} L(\omega-\nu) \, W(\nu) \, d\nu \qquad [3.10]$$

LOW-PASS FILTERS

where $W(\omega) = \mathcal{F}\{w(t)\}$. Let the transition band be $\Omega_1 \leq |\omega| \leq \Omega_2$ and let $L(\omega)$ be an ideal low-pass filter so that:

$$L(\omega) = L_1(\omega) = \begin{cases} 1 & \text{for } |\omega| \leq (\Omega_1 + \Omega_2)/2 \\ 0 & \text{otherwise} \end{cases}$$

and: $l_1(t) = \dfrac{1}{\pi t} \sin\left(\dfrac{\Omega_1 + \Omega_2}{2}\right) t$

The behaviour required of the approximation, $H_1(\omega)$, may be expressed as:

$$H_1(\omega) = \begin{cases} 1 & \text{for } |\omega| \leq \Omega_1 \\ 0 & \text{for } |\omega| > \Omega_2 \end{cases}$$

Further we require continuous roll-off for frequency interval $\Omega_1 \leq |\omega| \leq \Omega_2$ and the evenness property. From the definition of $L_1(\omega)$ it follows that:

$$L_1(\omega - \nu) = \begin{cases} 1 & \text{for } |\omega - \nu| \leq (\Omega_1 + \Omega_2)/2 \\ 0 & \text{for } |\omega - \nu| > (\Omega_1 + \Omega_2)/2 \end{cases}$$

Thus, the infinite integration limits in [3.10] may be substituted by $\omega \pm (\Omega_1 + \Omega_2)/2$ and the integrand itself may be simplified so that:

$$H_1(\omega) = \int_{\omega - (\Omega_1 + \Omega_2)/2}^{\omega + (\Omega_1 + \Omega_2)/2} W_1(\nu) \, d\nu \qquad [3.11]$$

To evaluate $H_1(\omega)$ for any particular frequency, $\omega = \Omega$, the integration is performed over the interval of length $\Omega_1 + \Omega_2$ with Ω as the center of the interval. The stop-band requirements for $H_1(\omega)$ are satisfied with any even function $W_1(\omega)$ which is zero for $|\omega| > (\Omega_1 + \Omega_2)/2$. In order to obtain unit gain in the pass band, $W_1(\omega)$ must have a unit area. A *boxcar function*, evidently the simplest form for $W_1(\omega)$, yields:

$$W_1(\omega) = \begin{cases} 1/\Delta\Omega & \text{for } |\omega| \leq \Delta\Omega/2 \\ 0 & \text{otherwise} \end{cases}$$

where $\Delta\Omega = \Omega_2 - \Omega_1$ and:

$$w_1(t) = \dfrac{1}{\pi t \Delta\Omega} \sin \dfrac{t\Delta\Omega}{2}$$

The integral in [3.11] provides linear roll-off so that the approximate frequency response function becomes (Fig. 3.2):

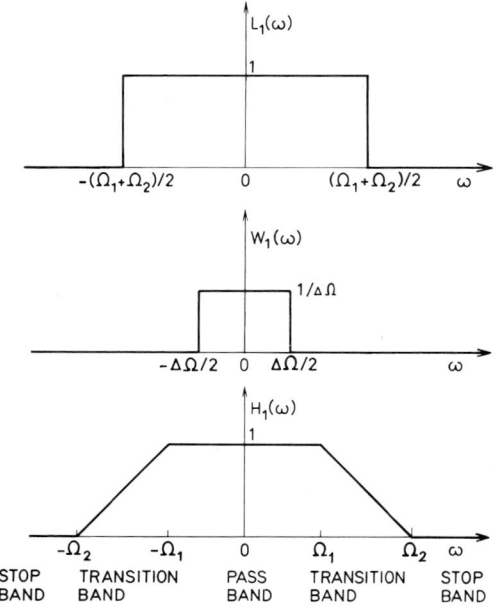

Fig. 3.2. Ideal transfer function, $L_1(\omega)$, frequency form of the weighting function, $W_1(\omega)$, and approximate transfer function, $H_1(\omega)$, of the Ormsby-type filter with a first-order roll-off.

$$H_1(\omega) = \begin{cases} 1 & \text{for } |\omega| \leq \Omega_1 \\ 0 & \text{for } |\omega| > \Omega_2 \\ (\omega + \Omega_2)/\Delta\Omega & \text{for } -\Omega_2 \leq \omega < -\Omega_1 \\ (\Omega_2 - \omega)/\Delta\Omega & \text{for } \Omega_1 \leq \omega < \Omega_2 \end{cases}$$

the impulse response $h_1(t)$ follows from [3.9] which gives:

$$h_1(t) = l_1(t)\, w_1(t) = \frac{1}{\pi t} \sin \frac{(\Omega_1+\Omega_2)t}{2} \times \frac{1}{\pi t \Delta\Omega} \sin \frac{t\Delta\Omega}{2} = \frac{\cos \Omega_1 t - \cos \Omega_2 t}{2\pi^2 \, t^2 \Delta\Omega} \qquad [3.12]$$

Quantizing $h_1(t)$ at equidistant time instants $t = n\Delta t$ where n again ranges over all integers from $-\infty$ to ∞, we obtain the impulse response of the corresponding approximate filter. Special care must be taken when evaluating h_n for $n = 0$. Functions $H_1(\omega)$ and/or $h_1(t)$ describe the so-called *Ormsby low-pass filter* with a first-order roll-off (Ormsby, 1961; Anders et al., 1964). Our result in [3.12] differs from that presented by Anders et al. (1964) by a factor of $1/2\pi$. This should be ascribed to different approaches, namely to cyclic and angular frequencies used by these authors and in our presentation, respectively.

Various weighting functions may be utilized in order to realize filters of specific

LOW-PASS FILTERS

roll-off in the transition bands. For example, *Martin-Graham filters* (Anders et al., 1964) make use of a cosine function so that $W(\omega)$ in [3.10] becomes:

$$W_2(\omega) = \begin{cases} \dfrac{\pi}{2\Delta\Omega} \cos \dfrac{\pi\omega}{\Delta\Omega} & \text{for } |\omega| \leq \Delta\Omega/2 \\ 0 & \text{otherwise} \end{cases}$$

The factor of $\pi/2\Delta\Omega$ has been introduced to normalize the area under $W_2(\omega)$ to unity. Since the weighting function $w_2(t)$ and $W_2(\omega)$ form a Fourier-transform pair we write:

$$w_2(t) = \frac{1}{2\pi} \int_{-\infty}^{\infty} \frac{\pi}{2\Delta\Omega} \cos \frac{\pi\omega}{\Delta\Omega} e^{j\omega t} d\omega =$$

$$= \frac{1}{2\Delta\Omega} \int_0^{\Delta\Omega/2} \cos \frac{\pi\omega}{\Delta\Omega} \cos \omega t \, d\omega = \frac{\cos(t\Delta\Omega/2)}{2\pi[1-(t\Delta\Omega/\pi)^2]}$$

As in the preceding case, the unit-impulse response of the approximate filter is defined as:

$$h_2(t) = l_2(t) w_2(t)$$

Let us suppose again that we wish to approximate a rectangular low-pass filter. Thus we set $l_2(t) = l_1(t)$ and the impulse response becomes:

$$h_2(t) = \frac{1}{2\pi^2 t} \sin \frac{(\Omega_1+\Omega_2)t}{2} \frac{\cos(t\Delta\Omega/2)}{1-(t\Delta\Omega/\pi)^2}$$

which after simplifying gives:

$$h_2(t) = \frac{\sin \Omega_2 t + \sin \Omega_1 t}{4t(\pi^2 - t^2 \Delta\Omega^2)} \qquad [3.13]$$

Function $h_2(t)$ has three singular points at $t = 0$, $\pm\Delta\omega/\pi$ which has to be carefully considered during the quantization. The transfer function of this type of Martin-Graham filter is determined from [3.11]. It is clear that for the frequency range $|\omega| < \Omega_1$, the total unit area beneath the cosine arc is to be taken into account which results in a uniform transfer function within the frequency range mentioned. For frequencies within the transition zone, only a certain fraction of the total cosine-arc area, depending upon that frequency, is included. Consequently, the roll-off is represented by a function which decreases monotonically from 1 for $\omega = \Omega_1$ to zero for $\omega = \Omega_2$. For $|\omega| > \Omega_2$ the transfer function is equal to zero since the entire cosine arc lies outside the integration interval (for details see Fig. 3.3). Summarizing, function $H_2(\omega)$ may be defined in the following way (Fig. 3.4):

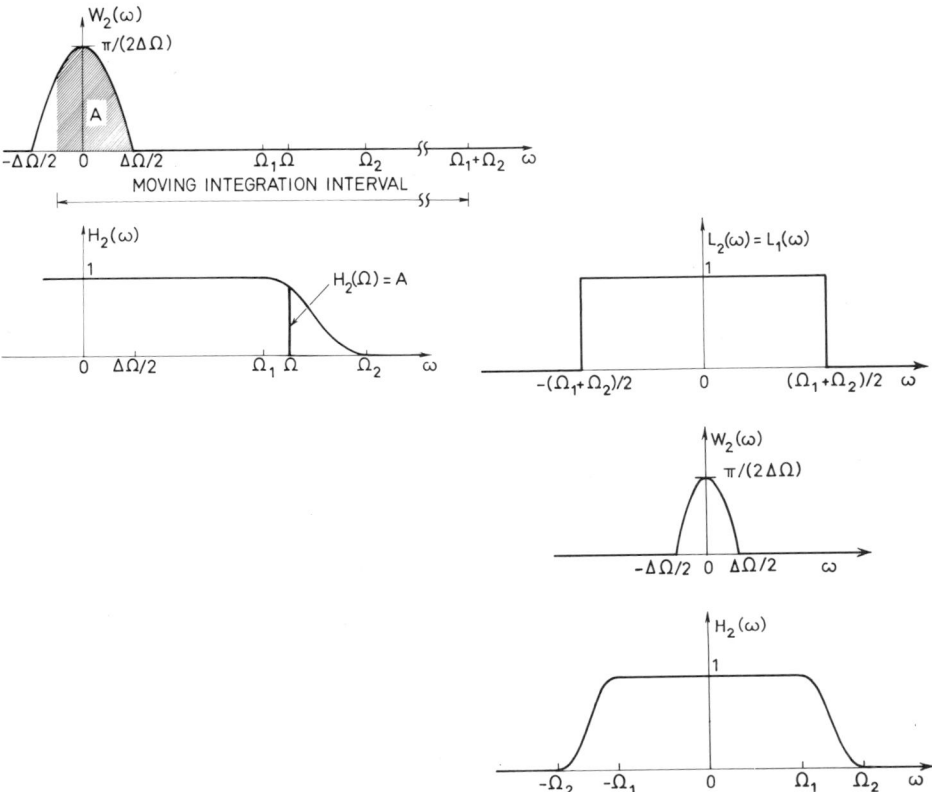

Fig. 3.3. Explanatory diagram to [3.12] for the case of Martin-Graham filters. Shaded region has and area A and Ω is the midpoint of the moving integration interval for the particular interval position depicted in the diagram. According to [3.12] the gain of the filter for $\omega = \Omega$, in the transition band, is equal to the shaded area.

Fig. 3.4. Ideal transfer function, $L_2(\omega)$, frequency form of the weighting function, $W_2(\omega)$, and approximate transfer function, $H_2(\omega)$, of the Martin-Graham filter with a cosine roll-off.

$$H_2(\omega) = \begin{cases} 1 & \text{for } |\omega| < \Omega_1 \\ \dfrac{1}{2}\left[1 + \cos\dfrac{\pi(\omega-\Omega_1)}{\Delta\Omega}\right] & \text{for } \Omega_1 \leq \omega \leq \Omega_2 \\ \dfrac{1}{2}\left[1 + \cos\dfrac{\pi(\omega+\Omega_1)}{\Delta\Omega}\right] & \text{for } -\Omega_2 \leq \omega \leq -\Omega_1 \\ 0 & \text{for } |\omega| > \Omega_2 \end{cases}$$

Transfer functions $H_1(\omega)$ and $H_2(\omega)$ are real and even functions of frequency.

This means that the Ormsby as well as the Martin-Graham filters discussed in this section perform a perfect phase-distortionless transmission.

By introducing quantized time into [3.12] and [3.13] we obtain power-series coefficients defining the system functions $H_1(z)$ and $H_2(z)$, respectively. Both $H_1(z)$ and $H_2(z)$ define nonrecursive filtering, which in the case of long input signals will be rather uneconomical. Due to the noncausal property either the data must be in a stored form, or delays described by [3.5] must be acceptable. As follows from the discussion in Section 1.2.1, the absolute integrability of the respective unit-impulse response is necessary to assure the filter stability. There are infinitely many functions applicable as $W(\omega)$. A reasonable choice may adequately approximate the desired filtering properties. Five different types of the roll-off together with the corresponding unit-impulse responses and transfer functions are listed by Anders et al. (1964); other examples may be found elsewhere.

3.1.4 *Butterworth filters*

Two examples of low-pass filters discussed in more detail in the previous section showed clearly the influence of amplitude-response discontinuities and of the impulse-response truncation upon oscillations of the amplitude characteristics. In this section we deal with another, qualitatively different approach to minimize the ripple of the amplitude response with the design of low-pass *Butterworth filters*. These filters approximate as closely as desired the behaviour of an ideal filter. The gain, $M_B(\omega)$, of these filters is monotonically decreasing with increasing frequency. The amplitude response does not have any ripple, either inside or outside the pass band. At frequency $\omega = 0$, Butterworth filters provide maximally flat amplitude response. Among further advantages belong the rather simple analytical expression for $M_B^2(\omega)$ and the easily controlled degree of attenuation in the transition band. Butterworth filters are usually specified by making use of their squared amplitude response, $M_B^2(\omega)$, which may be written as:

$$M_B^2(\omega) = |H_B(\omega)|^2 = \frac{1}{1 + (\omega/\omega_L)^{2n}}$$

where ω_L is the cut-off frequency and n is an integer defining the order of the filter. Here we see that for very low frequencies, $\omega \to 0$, the gain approaches unity. It is convenient to introduce a normalized frequency, $\nu = \omega/\omega_L$, so that the squared amplitude response becomes:

$$M_B^2(\omega) = \frac{1}{1 + \nu^{2n}} \qquad [3.14]$$

The amplitude response, $M_B(\omega)$, as a function of normalized frequency, ν, is diagrammed in Fig. 3.5 for the orders $n = 1$, 3 and 5. The low-pass properties of Butterworth filters are clear from this figure. Higher order n gives a better approximation of a rectangular response curve. At $\nu = 1$ the response $M_B(\omega)$ reaches a value of $1/\sqrt{2} = 0.707$ (about 3 dB attenuation) independently of n. The choice of n is usually ruled by attenuation requirements in the stop band.

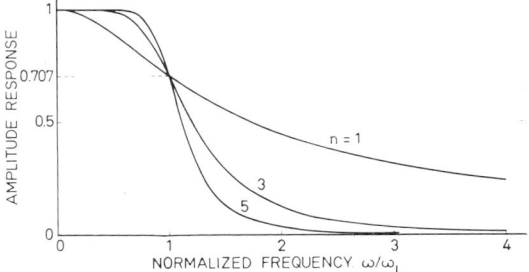

Fig. 3.5. Amplitude response, $M_B(\omega)$, vs. normalized frequency, ω/ω_L, of first-, third-, and fifth-order Butterworth filters.

Let us assume that we want to find the order of a low-pass Butterworth filter which is to have at least 10 dB attenuation at frequencies beyond $2\omega_L$. For a quick orientation, master curves like those presented in Fig. 3.5 may be used. A decrease of the amplitude level by 10 dB means a decrease to about 30% of the reference level at $\omega = 0$ which in our case corresponds to a decrease to amplitude equal to 0.3. From diagrams in Fig. 3.5 we read directly that a third-order filter would satisfy the attenuation requirements while a first-order filter would not. We check analytically whether the second-order filter (not shown in Fig. 3.5) could be used. Due to the decibel-scale definition we have:

$$20 \log [M(\nu_0)/M(\nu_2)] = 10 \log [M^2(\nu_0)/M^2(\nu_2)] \qquad [3.15]$$

where ν_0 is the normalized reference frequency, here $\nu_0 = 0$. Since we compare the amplitude at ν_0 with that at $\nu_2 = 2$, then from [3.14] and [3.15] it follows that:

$$10 \log \frac{1}{1/(1 + 2^{2n})} = 10 \log (1 + 2^{2n}) = 10$$

which gives:

$$1 + 2^{2n} = 10 \quad \text{and} \quad n \doteq 1.6$$

In other words, a Butterworth filter of order as low as 2 provides the required attenuation properties.

The next problem to solve consists of converting the frequency response, $H(\omega)$, of an analog filter into corresponding system function, $H(z)$, of a digital filter. One possible approach is to determine the poles of $H(\omega)$ and then apply one of the digitization techniques, described in Chapter 2, to obtain the corresponding digital representation. In terms of the Laplace transform, i.e. after introducing the complex frequency as the independent variable, the squared absolute value of the frequency response becomes:

LOW-PASS FILTERS

$$|H_B(p)|^2 = H_B^*(p)\,H_B(p) = \frac{1}{1+(-1)^n\,p^{2n}} \qquad [3.16]$$

where $p = \sigma + j\omega$ and $H_B^*(p)$ is the complex conjugate of $H_B(p)$. In [3.16] we assume normalized frequencies, otherwise p/ω_L should be used instead of p in the denominator on the right-hand side of the equation.

From [3.16] it follows that $H_B(p)\,H_B^*(p)$ has no finite zeros and the position of poles is determined from:

$$p^{2n} + (-1)^n = 0 \qquad [3.17]$$

As can be verified by complex-number theory, the roots of [3.17] in the p-plane are equally spaced around the unit circle with an angular separation of π/n. For n even the first pole occurs at an angle of $\pi/2n$ and for n odd the first pole occurs at zero angle. When the true frequency, ω, is used instead of the normalized frequency, ν, the poles are equally distributed in the p-plane on a circle of radius ω_L. Figure 3.6

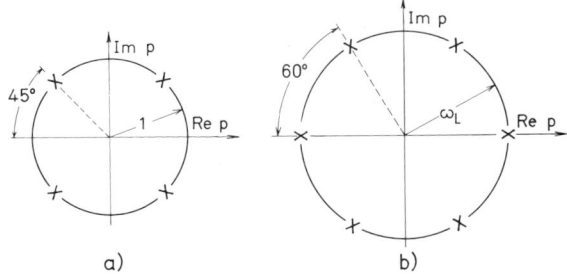

a)   b)

Fig. 3.6. Distribution of poles of Butterworth squared amplitude filter response in the p-plane: (a) $n = 2$, normalized frequency; (b) $n = 3$, true frequency.

shows pole distributions for the second- and third-order Butterworth filters with normalized and true frequencies, respectively. It follows directly from the figure that in the former case the poles are:

$$p_{1,2} = \cos 45° \pm j\sin 45° = 0.707 \pm j\,0.707$$
$$p_{3,4} = -\cos 45° \pm j\sin 45° = -0.707 \pm j\,0.707$$

whereas in the latter case we have:

$$p_5 = \omega_L$$
$$p_{6,7} = (0.5 \pm j\,0.866)\,\omega_L$$
$$p_{8,9} = (-0.5 \pm j\,0.866)\,\omega_L$$
$$p_{10} = -\omega_L$$

In a similar way, poles in the p-plane may easily be determined for any finite order n.

Perhaps it is worth mentioning that [3.14] and [3.16] provide formulae for the squared absolute value or the system function and not for the system function itself. Therefore, we do not need to worry about poles in the right half of the p-plane which otherwise would lead to an unstable system. In order to determine $H_B(p)$ we utilize the symmetry property of the pole distribution of $|H_B(p)|^2$. As follows from the two examples shown in Fig. 3.6, poles in the p-plane are located symmetrically about the real as well as about the imaginary axis. This property has a general validity irrespective of the filter order. Due to these properties, we may generate the function $H_B(p)$ by making use of only the poles located to the left from the imaginary axis, with the following results: firstly, $H_B(p)$ contains real and/or complex conjugate poles only, secondly, the pole distribution assures system stability and thirdly, the squared amplitude response is identical to that given by [3.14] or [3.16].

For the second- and third-order filters discussed above, the corresponding response functions, except for a constant multiplier, are:

$$H_2(p) = \frac{1}{(p-p_3)(p-p_4)} = \frac{1}{(p+0.707-j\,0.707)(p+0.707+j\,0.707)}$$

and:

$$H_3(p) = \frac{1}{(p+\omega_L)(p+0.5\,\omega_L-j\,0.866\,\omega_L)(p+0.5\,\omega_L+j\,0.866\,\omega_L)}$$

respectively. In order to ensure unit gain for frequencies $\omega \to 0$, functions $H_2(\omega)$ and $H_3(\omega)$ should be multiplied by $p_3 p_4$ and $p_8 p_9 p_{10}$, respectively. For the sake of simplicity, both these multiplications have been omitted. The pulse-transfer functions may be evaluated via the partial-fraction expansion. For $H_2(p)$ it is not necessary to carry out the expansion since the response function:

$$H_2(p) = \frac{1}{(p+0.707)^2+0.707^2}$$

may be found among those listed in Table II, so that:

$$H_2(z) = \frac{e^{-0.707\,\Delta t}\sin(0.707\Delta t)\,z^{-1}}{0.707\,[1-2e^{-0.707\,\Delta t}\cos(0.707\,\Delta t)\,z^{-1}+e^{-1.414\,\Delta t}\,z^{-2}]} \qquad [3.18]$$

The partial-fraction expansion transforms poles from the p-plane into corresponding poles in the z-plane; however, it does not preserve the location of zeros. Comparison of functions $H_2(p)$ and $H_2(z)$ shows that whereas the former function has no finite zeros the latter has a zero at $z^{-1}=0$. By using the geometrical z-plane representation of the pole-zero technique, it can be seen that this zero has no influence on the amplitude characteristics of the second-order Butterworth filter. In the case of higher-order filters, care must be taken with finite zeros introduced by the partial-fraction expansion. For sufficiently dense digitizing these zeros usually have negligible effect in the pass band. For more detail the reader is referred to Gold and Rader (1969).

LOW-PASS FILTERS

Another possibility to determine the digital representation of $H(p)$ is the application of the bilinear transformation. Let us assume that a second-order filter satisfies the attenuation requirements, and that the sampling period and the cut-off frequency are $\Delta t = 0.1$ sec and $\omega_L = 4\pi$ rad/sec, respectively. According to the instruction presented in Section 2.5.3 we first transform the cut-off frequency into the fictitious-frequency domain. From [2.38] it follows that:

$$\omega_A = \tan(\omega_L \Delta t/2) = \tan(0.4\pi/2) = 0.727 \text{ rad/sec}$$

The corresponding second-order Butterworth filter has a pair of complex-conjugate poles $p_{1,2} = (-0.707 \pm j0.707) \, 0.727$. Neglecting the gain factor, the transfer function in the fictitious-frequency domain becomes:

$$H(p) = \frac{1}{p^2 + 1.028 p + 0.528}$$

In order to determine the digital representation, $H(z)$, we simply replace p in $H(p)$ by $(1 - z^{-1})/(1 + z^{-1})$, which gives:

$$H(z) = \frac{1}{[(1 - z^{-1})/(1 + z^{-1})]^2 + 1.028 (1 - z^{-1})/(1 + z^{-1}) + 0.528}$$

$$= 0.391 \frac{1 + 2z^{-1} + z^{-2}}{1 + 0.022z^{-1} - 0.196z^{-2}}$$

The recursion relation for the nth output term is:

$$y_n = 0.391x_n + 0.782x_{n-1} + 0.391x_{n-2} - 0.022y_{n-1} + 0.196y_{n-2}$$

Should it be necessary to preserve unit gain at $\omega = 0$, $H(z)$ has to be multiplied by a factor of 0.528 (note that for $\omega = 0$ the variable $z = e^{j\omega\Delta t} = 1$).

The bilinear transformation may be applied directly to the squared amplitude response, $M_B^2(\omega)$. After prewarping the continuous cut-off frequency and substituting $p = (1 - z^{-1})/(1 + z^{-1})$, [3.16] gives:

$$|H_{2n}(z)|^2 = \frac{1}{1 + (-1)^n [(1 - z^{-1})/(1 + z^{-1})]^{2n}/\tan^{2n}(\omega_L \Delta t/2)}$$

$$= \frac{\tan^{2n}(\omega_L \Delta t/2)}{\tan^{2n}(\omega_L \Delta t/2) + (-1)^n [(1 - z^{-1})/(1 + z^{-1})]^{2n}} \quad [3.19]$$

which will contain stable as well as unstable poles. Repeating the procedure utilized in the p-plane we determine the stable system function, $H_n(z)$, of the desired Butterworth filter. Separation of stable and unstable poles in $|H_{2n}(z)|^2$ makes it possible to construct $H_n(z)$ by including all the stable poles of $|H_{2n}(z)|^2$ and excluding the unstable poles.

Equation [3.19] is also interesting from another point of view. Introducing the relation $z = e^{j\omega\Delta t}$ into the ratio $(1 - z^{-1})/(1 + z^{-1})$ we find:

$$\left[\frac{1-z^{-1}}{1+z^{-1}}\right]_{z=e^{j\omega\Delta t}} = \frac{1-e^{-j\omega\Delta t}}{1+e^{-j\omega\Delta t}} = \tanh(j\omega\Delta t/2) = j\tan(\omega\Delta t/2)$$

and [3.19] may be rewritten as:

$$|H_{2n}(\omega)|^2 = \frac{1}{1+[\tan(\omega\Delta t/2)/\tan(\omega_L\Delta t/2)]^{2n}} \qquad [3.20]$$

Filters with the squared amplitude response given by [3.20], called *tangent filters* belong to a broader class of so-called *trigonometric* (sine, cosine, tangent) *filters*. The direct relation of tangent filters to the bilinear transformation follows from [3.19]. The digital version of the tangent filter is given by the right-hand side of [3.19] provided that only stable poles are included. The distribution of poles and zeros of $|H_{2n}(z)|^2$ in the z-plane is discussed in detail by Gold and Rader (1969) and Ackroyd (1973). Characteristics of several basic trigonometric filters have been listed by Holtz and Leondes (1966) and an extensive discussion of tangent and sine filters may be found in Otnes and Enochson (1972). Butterworth filters have been applied e.g. by Floyd (1969) to reduce the noise effect in gravity measurements.

3.1.5 Chebyshev filters

Another category of low-pass filters which may be defined, except for the constant gain factor, by the location of their poles are the low-pass *Chebyshev filters*. In general, the performance of a Chebyshev filter is given, as in the case of Butterworth filter, by its squared amplitude response:

$$M_C^2(\omega) = |H_C(\omega)|^2 = \frac{1}{1+\epsilon^2 C_N^2(\omega/\omega_L)} \qquad [3.21]$$

Equation [3.21] defines a Chebyshev filter with $M_C^2(\omega)$ which has uniform ripple in the pass band and is monotonically decreasing in the stop band. The parameter ϵ controls the ripple amplitude in the pass band. The peak-to-peak ripple amplitude, δ_A, is evaluated from the formula (see e.g. Rader and Gold, 1967 or Rabiner et al., 1972):

$$\delta_A = 1 - (1+\epsilon^2)^{-1/2}$$

$C_N(\omega)$ is a *Chebyshev polynomial* of degree N. These polynomials may be generated by making use of a three-term recursion relationship (see e.g. Hamming, 1962, p. 251):

$$C_0(\omega) = 1$$
$$C_1(\omega) = \omega$$
$$C_{N+1}(\omega) + C_{N-1}(\omega) = 2\omega\, C_N(\omega) \qquad \text{for } N \geq 1$$

LOW-PASS FILTERS

As an example, consider $N = 3$ and $\epsilon_A = 0.6$ ($\delta_A = 0.14$). Within the pass band the amplitude response of the filter will undulate between values of 0.86 and 1 thus corresponding to approximately 1.3 dB ripple. The characteristics $M_C(\omega)$ for $N = 3$ and $\epsilon = 0.6$ are shown in Fig. 3.7 where the third-order Butterworth

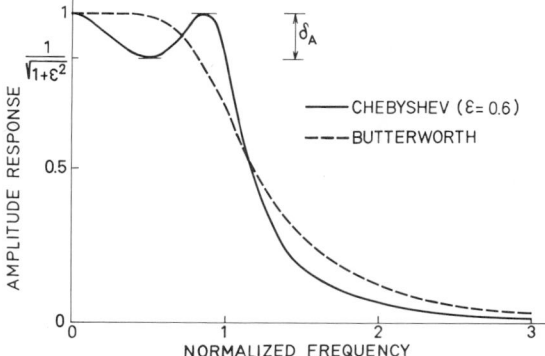

Fig. 3.7. Comparison of amplitude responses of third-order Chebyshev and Butterworth filters. $\delta_A = 0.14$ is the peak-to-peak amplitude of the ripple.

response is also diagrammed for comparison. The figure makes it clear that a Chebyshev filter has a steeper roll-off than a Butterworth filter of the same order. The price to be paid for this improvement is the ripple in the pass band. It is worth mentioning that for N odd and N even the amplitude response starts at $\omega = 0$ from values 1 and $1 - \delta_A$, respectively.

In order to construct the digital representation of a Chebyshev filter, poles of $M_C^2(\omega)$, i.e. roots of the denominator of the right-hand side of [3.21] must be determined. In the p-plane, the poles are again distributed symmetrically about both the real and imaginary axis, thus providing a mixture of stable and unstable poles. To ensure a stable filter, the response function, $H_C(\omega)$, must include all of the poles from the left half of the p-plane only. Applying any of the digitizing methods, such as the bilinear transformation, to $H_C(\omega)$, the system function $H(z)$ is found.

Rader and Gold (1967) present a rather simple method for determining the poles of $M_C^2(\omega)$ for given parameters ϵ, N and ω_L. Poles are located along an ellipse (see Fig. 3.8) with vertical and horizontal axes equal to $2b\omega_L$ and $2a\omega_L$, respectively where:

$$b, a = [(\sqrt{\epsilon^2+1}+\epsilon^{-1})^{1/N} \pm (\sqrt{\epsilon^2+1}+\epsilon^{-1})^{-1/N}]/2$$

Figure 3.8 shows the position of poles on the ellipse for a third-order Chebyshev filter, determined by utilizing two Butterworth circles of diameters $2a\omega_L$ and $2b\omega_L$.

As with the Butterworth low-pass filters, several types of trigonometric Cheby-

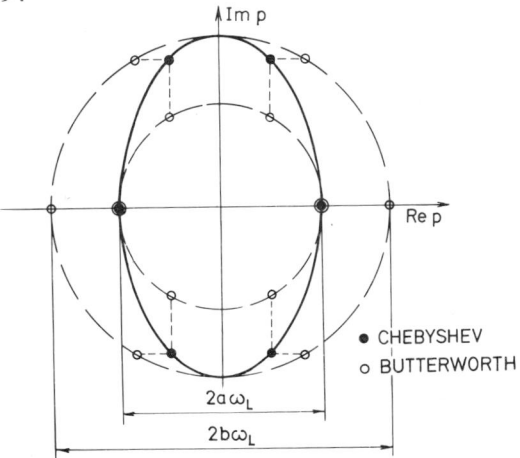

Fig. 3.8. Geometrical determination of poles of a third-order Chebyshev low-pass filter. Two Butterworth circles of diameters $2a\omega_L$ and $2b\omega_L$ are used in the determination.

shev filters may be constructed. For example a *tangent Chebyshev* low-pass *filter* has the form (see e.g. Rader and Gold, 1967; Ackroyd, 1973):

$$|H(\omega)|^2 = \frac{1}{1 + \epsilon^2 \, C_N^2 \, [\tan (\omega \Delta t/2)/\tan (\omega_L \Delta t/2)]}$$

Other forms may be found in Otnes and Enochson (1972) or elsewhere.

3.2 HIGH-PASS FILTERS

Following the approach outlined in Section 3.1, we may also look at the characteristics of high-pass filters. Because of reasons already discussed, we introduce negative frequencies so that the amplitude response, $M(\omega)$, becomes an even function. If we limit ourselves to frequencies $|\omega| \leq \omega_N = \omega_s/2$ and consider a zero-phase shift, an ideal high-pass filter is defined by:

$$H(\omega) = M(\omega) = \begin{cases} 0 & \text{for } |\omega| < \omega_H \\ A & \text{for } |\omega| \geq \omega_H \end{cases} \qquad [3.22]$$

where again A is a real constant defining the gain of the filter in the pass band. The analog impulse response is derived directly from [3.2] by substituting the appropriate integration limits, so that:

$$h(t) = \frac{A}{\pi} \int_{\omega_H}^{\omega_N} \cos \omega t \, d\omega = \frac{A}{\pi t} [\sin \omega_N t - \sin \omega_H t] \qquad [3.23]$$

Introducing quantized time $t = n\Delta t$ and $\omega_N = \pi/\Delta t$, the first term in the brackets

HIGH-PASS FILTERS

becomes zero for any integer n and hence the coefficients of the impulse-response sequence are:

$$h_n = -\frac{A\omega_H}{\pi} \frac{\sin \omega_H n\Delta t}{\omega_H n\Delta t} \quad \text{for } |n| = 1, 2, \ldots$$

Care must be taken when evaluating $h(t)$ for $t = 0$. Using l'Hospital's rule we obtain:

$$h(0) = h_0 = A(\omega_N - \omega_H)/\pi$$

Similarity between [3.3] and [3.23] makes it easy to determine values of h_n, $|n| = 1, 2, \ldots$, of a high-pass filter in terms of weights of a corresponding low-pass filter. Determine a low-pass filter impulse-response sequence, assuming a cut-off frequency ω_H, and change the sign of all the weights. As far as concerns the value of h_0 the formula derived above has to be utilized. This approach, carried out entirely in the time domain, is convenient for generating the nonrecursive high-pass filters, provided that the weights of the corresponding low-pass filter are known.

Even in the frequency domain, the generation of high-pass filters can be based upon the low-pass filter design. For example, a reasonable high-pass Butterworth filter may be defined in terms of its squared amplitude response as:

$$M_H^2(\omega) = 1 - M_L^2(\omega) \quad [3.24]$$

where $M_H^2(\omega)$ and $M_L^2(\omega)$ are squared amplitude responses of the high- and low-pass filter, respectively and $\omega_L = \omega_H$. Thus, a high-pass version of the Butterworth filter can be written as:

$$M_H^2(\omega) = 1 - \frac{1}{1 + (\omega/\omega_H)^{2n}} = \frac{(\omega/\omega_H)^{2n}}{1 + (\omega/\omega_H)^{2n}} \quad [3.25]$$

Comparing the forms of $M_H^2(\omega)$ and $M_L^2(\omega)$ it is evident that $M_H^2(\omega)$ has a $2n$th-order zero at $\omega = 0$, whereas $M_L^2(\omega)$ has no finite zero. By contrast, both functions have the same pole distribution. In order to carry out the conversion from analog- into digital-filter representation techniques discussed in Section 3.1.4 are applicable.

An alternative possibility of converting the low-pass filter into the corresponding high-pass filter is to apply a suitable frequency transformation. For example, the transformation $\omega = 1/v$ provides a correct solution. Substitution of $\omega = 1/v$ and $\omega_H = \omega_L = 1/v_L$ into the Butterworth squared amplitude response gives:

$$M_H^2(v) = M_L^2\left(\omega = \frac{1}{v}\right) = \frac{(v/v_L)^{2n}}{1 + (v/v_L)^{2n}}$$

and the resemblance with [3.25] is evident. The low-pass to high-pass conversion may be carried out directly in the z-domain. Replacing z by $-z$ in the original low-pass system function, we obtain a high-pass filter. If the original cut-off frequency was f_L then the new cut-off frequency becomes $f_H = 1/2\Delta t - f_L$ (see Ackroyd, 1973). Figure 3.9 shows the amplitude response of a third-order high-

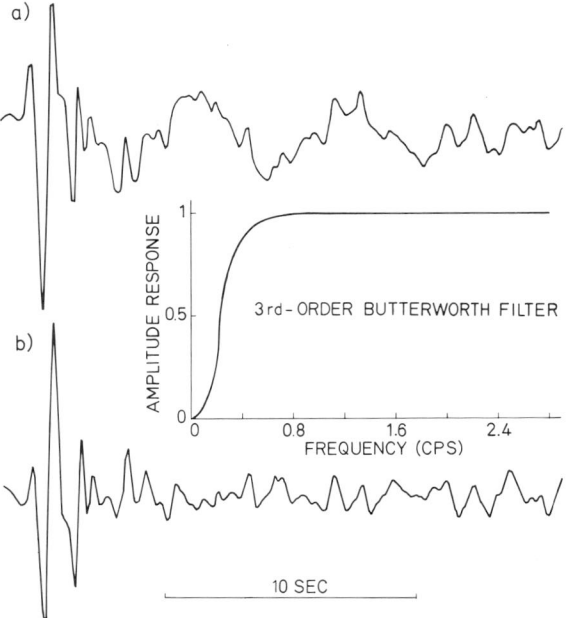

Fig. 3.9. Amplitude response of a third-order Butterworth high-pass filter together with the (a) original; (b) filtered seismograms. After J. S. Farnbach (personal communication, 1974).

pass Butterworth filter calculated via the bilinear transformation. The filter has been applied to a seismogram (a) of a teleseismic event in order to suppress the noise with the dominant period of about 6 sec. The efficacy of the filter is seen in the filtered trace (b) where the undesired noise is no more visible.

Generally, given a low-pass filter response function, $H_L(\omega)$, the selectivity property of a corresponding high-pass filter, $H_H(\omega)$, may be expressed as:

$$|H_H(\omega)|^2 = 1 - |H_L(\omega)|^2$$

provided that $|H_L(\omega)|$ is a monotonic function of ω. As an application of high-pass filtering by means of low-pass filters the barometric pressure high-pass filtering as presented by Holloway (1958) can be cited.

3.3 BAND-PASS FILTERS

Generation of band-pass filters may be significantly simplified by taking advantage of results available from the design of low-pass filters outlined in

Section 3.1. Several design possibilities are discussed below. In order to construct weighting coefficients assuring real and phase-distortionless zero-delay output, only filters with real and even frequency response will be considered.

Let us assume an ideal symmetric low-pass filter with the cut-off frequency, ω_L, whose frequency response, $H_L(\omega)$, has a unit gain in the band $|\omega| \leq \omega_L$. Simple shifting of $H_L(\omega)$ to the right by an amount of $\omega_0 > \omega_L$ gives a band-pass filter characteristics with a *bandwidth* $2\omega_L$ and *center frequency* ω_0. To preserve the evenness property we divide $H_L(\omega)$ into two identical characteristics each of them having the pass-band gain equal to 1/2. We shift $H_L(\omega)/2$ to the right by ω_0 and obtain, due to [3.2] and to the frequency-shifting theorem, the impulse response of a band-pass filter in a form:

$$h_1(t) = [\exp(j\omega_0 t) \sin \omega_L t]/2\pi t$$

Similarly, we shift $H_L(\omega)/2$ also to the left by the same amount, ω_0, and obtain:

$$h_2(t) = [\exp(-j\omega_0 t) \sin \omega_L t]/2\pi t$$

Evidently, the unit-impulse response of a symmetric band-pass filter is given by the sum:

$$h(t) = h_1(t) + h_2(t) = \frac{\omega_L}{\pi} \frac{\sin \omega_L t}{\omega_L t} \cos \omega_0 t \qquad [3.26]$$

where ω_L is one half of the filter bandwidth and ω_0 is the center frequency. The band-pass filter described by [3.26] has been used e.g. by Crampin and Båth (1965) for mode separation of seismic surface waves and for suppressing the microseisms on seismograms (see also Landisman et al., 1969 or Herrmann, 1973). To minimize the truncation effect, a triangular-shaped time window has been utilized. Mode separation of surface waves by means of low-pass filters has been carried out by Savarenskiĭ and Kosarev (1967). Third-order Butterworth band-pass filter with corner frequencies 1.2 cps and 3.2 cps has been employed at the Norwegian Seismic Array (NORSAR) to suppress the undesired noise (Gjøystdal and Husebye, 1972).

Substituting [3.2] into [3.26] and assuming a unit gain in the pass band we have:

$$h(t) = 2h_L(t) \cos \omega_0 t \qquad [3.27]$$

where $h_L(t)$ is the impulse response of the so-called *equivalent low-pass filter* defined by [3.2]. The impulse response, $h(t)$, given by [3.27] can be viewed as an amplitude-modulated signal with an envelope $2h_L(t)$ and carrier frequency ω_0.

Band-pass filters may be also realized as the subtraction of one low-pass filter, $H_1(\omega)$, from another, $H_2(\omega)$. Suppose that the cut-off frequencies are ω_1 and ω_2, respectively, where $\omega_1 < \omega_2$ and that both filters have unit gain in the pass band. Then, the response becomes (see e.g. also Zelei, 1971):

$$H(\omega) = H_2(\omega) - H_1(\omega) = \begin{cases} 1 & \text{for } \omega_1 \leq |\omega| \leq \omega_2 \\ 0 & \text{elsewhere} \end{cases}$$

representing an ideal rectangular band-pass filter. The impulse response, $h(t)$, is determined in terms of the two low-pass filters as:

$$h(t) = h_2(t) - h_1(t) = (\sin \omega_2 t - \sin \omega_1 t)/\pi t \qquad [3.28]$$

If we realize that $(\omega_1 + \omega_2)/2$ gives the center frequency and $(\omega_2 - \omega_1)/2$ is equal to one half of the filter bandwidth, the identity of [3.26] and [3.28] follows from well-known trigonometric relations.

The shifting or subtraction procedures described above are generally useful in designing convolution band-pass filters. When constructing recursive filters, we take the advantage of the convolution theorem. Any band-pass filter may be replaced by a serial arrangement of appropriate low- and high-pass filters. An input signal enters first, say, the low-pass filter which cuts off the prescribed part of its high-frequency components. The remaining portion of the signal is further fed into the high-pass filter which eliminates the low-frequency part of the input spectrum. Hence, the final output contains only components within a certain frequency interval, specified by the cut-off frequencies of both filters and the respective roll-off. Since multiplication in the frequency domain corresponds to convolution in the time domain, the resulting band-pass filter, $H(\omega)$, is obtained as a product of the low-pass filter, $H_L(\omega)$, and the high-pass filter, $H_H(\omega)$. These filtering procedures are linear operations so that the order of the low- and corresponding high-pass filter does not affect the final result.

Mooney (1968) presents a simplified pole-zero technique which makes it possible to design low- and high-pass filters as well as band-pass and band-rejection filters. The method consists in locating the poles and zeros in the z-plane (see also Section 2.4) so as to obtain the desired frequency characteristics. After certain experience and with the help of master curves the location of poles and zeros can be made more or less intuitively and so Mooney's method usually provides a quick solution. Since the design is carried out directly in the z-plane, the analog–digital conversion is avoided. A similar approach applied to band-rejection filters has been used by Göncz and Zelei (1972).

Chapter 4

CORRELATION AND OPTIMUM FILTERS

In this chapter our primary concern will be the application of second-order statistical characteristics, i.e. correlation functions and power spectra of the signal and noise, in the synthesis of filters for efficient signal–noise discrimination. It will be shown that under certain assumptions a direct utilization of statistical characteristics may be viewed as an alternative filtering process.

A large part of the chapter is devoted to the design of *optimum filters*, i.e. filters which perform the best separation of the signal and noise in a given sense. There are a number of performance criteria to define various optimum filters, but we shall focus our attention on two types, the *matched* and *Wiener optimum filters*.

The last section of this chapter discusses briefly *polarization filters* as a special case of *correlation filters*.

Design principles are illustrated by making use of several typical geophysical applications. Detailed mathematical derivations, minimized here, are available in Davenport and Root (1958), Helstrom (1960), Lee (1960), Solodovnikov (1960) or elsewhere. An extensive list of applications from various scientific branches has been accumulated e.g. by Anstey (1964).

4.1 CORRELATION FILTERS

In some applications it may be required to answer the question whether or not a periodic component, buried in the background noise, exists in the recorded trace $x(t)$. Such a problem may be considered as a typical one to be solved by making use of autocorrelation functions. Under certain assumptions aperiodic signals may also be enhanced by utilizing the cross-correlation.

4.1.1 *Detection of periodic signals by autocorrelation*

Let us start with a simple case, that of a trace which is a periodic function of time with a period T, so that $x(t) = x(t + nT)$ where $n = 0, \pm 1, \pm 2, \ldots$. The Fourier expansion of the trace is:

$$x(t) = \frac{a_0}{2} + \sum_{n=1}^{\infty} (a_n \cos n\Omega t + b_n \sin n\Omega t)$$

where the fundamental frequency is $\Omega = 2\pi/T$. The autocorrelation function, $R_{xx}(\tau)$, of the recorded trace may be expressed in terms of the coefficients of the Fourier expansion (see e.g. Lee, 1960, pp. 11–14 or Solodovnikov, 1960, p. 93) as:

$$R_{xx}(\tau) = \frac{a_0^2}{4} + \frac{1}{2} \sum_{n=1}^{\infty} (a_n^2 + b_n^2) \cos n\Omega\tau \qquad [4.1]$$

In other words, the autocorrelation function of a periodic trace is again a periodic function with the original fundamental frequency. $R_{xx}(\tau)$ contains all frequency components existing in $x(t)$ without phase shift and with amplitudes equal to the mean-square values of the corresponding Fourier coefficients. Thus, if the trace is a pure sine wave, $x(t) = A \sin(\Omega t + \phi)$, the oscillations in $R_{xx}(\tau)$ would have amplitudes equal to $A^2/2$, which means an amplitude change by a factor of $A/2$. Since the amplitude change depends upon the amplitude itself, autocorrelation is a nonlinear function.

So far we have assumed the recorded trace to be a periodic function of time. In practice, however, this would be rather an exceptional case. A common model in processing seismic data is, broadly speaking, the following. Seismic waves propagate from the focus through various parts of the earth's interior and are recorded on or close to the earth's surface together with various kinds of noise. The earth's interior changes the spectrum of signals, as they propagate from the source to the receiver, in a similar way as discussed for filters (see e.g. Backus, 1959; Lindsey, 1960; Sengbush et al., 1961; Treitel and Robinson, 1966a). The above model is not limited to seismic problems but may be utilized in other geophysical fields as well (see e.g. George et al., 1964).

In accordance with this model, consider a more practical case, namely a trace consisting of a periodic signal $s(t)$ and a random noise $n(t)$:

$$x(t) = s(t) + n(t)$$

Consider an aperiodic noise which has zero mean value and is not correlated with $s(t)$. The trace-autocorrelation function is:

$$\begin{aligned}R_{xx}(\tau) &= \lim_{T_1 \to \infty} \frac{1}{2T_1} \int_{-T_1}^{T_1} x(t) \, x(t+\tau) \, dt = \\ &= \lim_{T_1 \to \infty} \frac{1}{2T_1} \int_{-T_1}^{T_1} [s(t) + n(t)] \, [s(t+\tau) + n(t+\tau)] \, dt = \\ &= R_{ss}(\tau) + R_{nn}(\tau) + R_{sn}(\tau) + R_{ns}(\tau) \qquad [4.2]\end{aligned}$$

Since $s(t)$ and $n(t)$ are uncorrelated, the last two terms on the right-hand side of [4.2] are zero. Details depend upon spectral properties of the noise, nevertheless for aperiodic noise $R_{nn}(\tau)$ decreases with increasing τ. This means that beyond a certain finite value, $\tau = \tau_1$, $R_{nn}(\tau)$ may be neglected in [4.2].

For $\tau > \tau_1$ the trace-autocorrelation function may be approximated simply by $R_{ss}(\tau)$. Due to [4.1], for large enough τ, the function $R_{xx}(\tau)$ will show a periodicity similar to that of the periodic part of the recorded trace, $x(t)$. Hence, the trace autocorrelation reveals the existence of $s(t)$ in the mixture, $x(t)$. $R_{xx}(\tau)$ does not resemble the form of $s(t)$, but the main advantage of this type of detection is that it is not necessary that the period of $s(t)$ be known a priori.

When processing seismic signals, the case may be further complicated by the fact that also the noise, e.g. microseisms, contains a periodic component. The applicability of the method then depends upon the signal power relative to that of the noise. Figure 4.1 shows portions of normalized autocorrelation functions of seismic records of three aftershocks of a Kurile Islands earthquake with different signal/noise ratios. The autocorrelation function in diagram 4.1a (low signal/noise ratio) contains clear low- as well as high-frequency components. The former being due to the microseismic noise with a predominant period of about 7 sec while the latter, with a period of about 1 sec, belongs to the recorded aftershock. In diagram 4.1b (medium signal/noise ratio) the long-period component is no longer as evident as in the preceding case. In the autocorrelation in Fig. 4.1c (high signal/noise ratio) the noise component is too small to be identified visually. As can be seen from all three diagrams, the autocorrelation function does not resemble the form of the original trace but evidently enhances the existing periodicity. The autocorrelation technique has been utilized e.g. by Mikulski and Mikulska (1973) in investigating the periodicity of hydrometeorological phenomena.

4.1.2 Detection of signals by cross-correlation

The cross-correlation function of two waveforms is a measure of the resemblance of the waveforms as a function of their relative time shift. Cross-correlating two traces of similar form results in high cross-correlation whereas traces of different

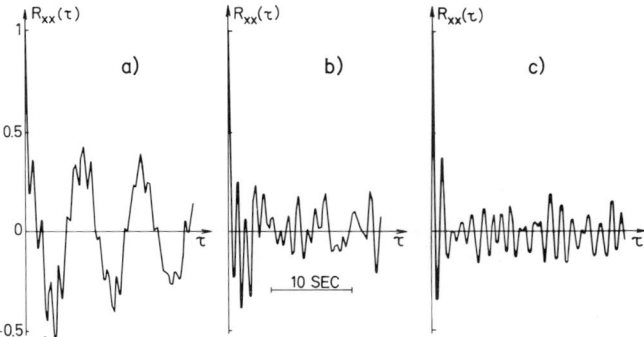

Fig. 4.1. Normalized autocorrelation functions of a mixture of 7 sec microseisms with seismic signals from three different aftershocks of a Kurile Islands earthquake with: (a) low; (b) medium; (c) high signal/noise ratio.

form provide low cross-correlation. This essential behaviour may be employed in detection of signals buried in noise.

Due to our knowledge and experience, some basic characteristics, especially predominant periods, of recorded signals may at times be predicted within certain limits. If this is the case, then signal detection by means of cross-correlation can be viewed as a search of the trace for arrivals of the expected form. We may say that the process is searching for a copy of the expected signal in the recorded trace. Obviously, the success of the cross-correlation technique depends above all upon the extent to which the recorded signals may be predicted.

For instance, if the recorded trace contains a periodic signal then in the simplest case cross-correlating the trace against a sinusoid with the signal period would extract the signal from the random noise. The cross-correlation function in this case consists only of components close to the frequency of the sine wave. When the expected signal is an infinitely long sine wave, the cross-correlation corresponds to filtering by an infinitely narrow band-pass filter. The similarity between cross-correlation and filtering operations is an important phenomenon frequently utilized in data processing.

Let us investigate in more detail some useful relations in both the time and frequency domains. Consider two periodic time functions, $x_1(t)$ and $x_2(t)$, with the same fundamental frequency $\Omega = 2\pi/T$. Fourier expansion of these functions yields:

$$x_1(t) = \frac{a_{10}}{2} + \sum_{n=1}^{\infty} (a_{1n} \cos n\Omega t + b_{1n} \sin n\Omega t)$$

$$x_2(t) = \frac{a_{20}}{2} + \sum_{n=1}^{\infty} (a_{2n} \cos n\Omega t + b_{2n} \sin n\Omega t)$$

According to Lee (1960, pp. 19–20) the corresponding cross-correlation function, $R_{12}(\tau)$, may be expressed by means of the Fourier-expansion coefficients, similarly to the case of autocorrelation, as:

$$R_{12}(\tau) = \frac{a_{10} a_{20}}{4} + \tfrac{1}{2} \sum_{n=1}^{\infty} [(a_{1n}^2 + b_{1n}^2)(a_{2n}^2 + b_{2n}^2)]^{1/2} \cos(n\Omega\tau + \theta_{2n} - \theta_{1n}) \qquad [4.3]$$

where $\theta_n = \tan^{-1}(-b_n/a_n)$. For example, cross-correlating two stationary sine waves with the same frequency, Ω, and with amplitudes A_1 and A_2 results in a cosine cross-correlation function with a period Ω and a changed amplitude $A_1 A_2/2$. Note that the zero-frequency as well as the fundamental and all higher harmonic coefficients in [4.3] appear as products of corresponding Fourier coefficients. This means that if $x_1(t)$ or $x_2(t)$ does not contain a particular component, the corresponding component will also be absent in $R_{12}(\tau)$. In other words, the cross-correlation function of two periodic functions is again periodic and contains only frequencies common to both functions. It follows from [4.3] that, unlike the autocorrelation defined by [4.1], the function $R_{12}(\tau)$ retains the phase differences

$\theta_{2n} - \theta_{1n}$. An application of the cross-correlation technique to seismic data may be found e.g. in Merkel and Alexander (1969). Bhimasankaram et al. (1973) employed interprofile cross-correlation to reveal hidden anomalies in magnetic data from an areal survey. In order to answer the question whether observed gravity and magnetic anomalies are generated by the same geological body, Botezatu and Calota (1973) employed the correlation technique. They found that the cross-correlation between the curves of gravity and magnetic anomalies, measured along the same profile, is a powerful tool in solving this problem.

Turning to a more general case, let us assume the recorded trace, $x(t) = s(t) + n(t)$, to be a mixture of a periodic signal and random noise as was the case above. The period of the signal is known or has been determined by taking the autocorrelation. Cross-correlation of $x(t)$ against a periodic signal $v(t)$ which has the same period as the signal $s(t)$ yields:

$$R_{xv}(\tau) = R_{sv}(\tau) + R_{nv}(\tau) \qquad [4.4]$$

The term $R_{nv}(\tau)$, being the cross-correlation of two dissimilar waveforms, can be neglected and $R_{xv}(\tau)$ becomes:

$$R_{xv}(\tau) \simeq R_{sv}(\tau)$$

Function $R_{xv}(\tau)$ contains only frequencies common to $s(t)$ and $v(t)$ and so the cross-correlation can be viewed as a general filtering process.

To demonstrate the analogy between cross-correlation and filtering we investigate the appropriate relations in the frequency domain. Consider a linear system with input $x(t)$ and impulse response $h(t)$. According to the definition (see e.g. Solodovnikov, 1960, p. 97) the cross-power spectrum $\Phi_{hx}(\omega)$, is:

$$\Phi_{hx}(\omega) = H^*(\omega) X(\omega) \qquad [4.5]$$

where $H(\omega) = \mathscr{F}\{h(t)\}$ and $X(\omega) = \mathscr{F}\{x(t)\}$. The expression $H^*(\omega) = \mathscr{F}\{h(-t)\}$ denotes the complex conjugate of $H(\omega)$. Comparison of [4.5] and [1.14] shows that correlating an arbitrary waveform $h(t)$ with an arbitrary input $x(t)$ is equivalent to transmitting $x(t)$ through a filter whose amplitude response is the amplitude spectrum of $h(t)$, and whose phase response is the negative of the phase spectrum of $h(t)$.

In general, recorded geophysical signals occupy a certain frequency range rather than only one exact frequency from the whole frequency spectrum. Therefore, the cross-correlation with a pure sine wave as mentioned above does not necessarily provide satisfactory means for signal detection. Significant improvement may be achieved by making use of waveforms with a linearly swept frequency, so-called *chirp filtering*. For example a sample of time duration of L sec with linearly varying frequency, also called a *chirp waveform*, can be written in a form:

$$h(t) = \begin{cases} \sin \theta(t) = \sin 2\pi [f_0 + (f_1 - f_0)t/(2L)]t & \text{for } 0 \leq t \leq L \\ 0 & \text{otherwise} \end{cases}$$

where f_0 and f_1 are the initial and final cyclic frequencies of the sample, respectively. The instantaneous frequency, at time $t = t_1$, within the interval considered, is defined as the time derivative:

$$\left[\frac{d\theta(t)}{dt}\right]_{t=t_1}.$$

Figure 4.2 depicts an example of $h(t)$ with parameter values $L = 20$ sec, $f_0 = 1$ cps and $f_1 = 3$ cps. It can be seen in the upper part of the figure that the frequency of the waveform varies linearly from 1 cps at the beginning to 3 cps at the end of the sample. Due to the finite length of the sample, the amplitude spectrum of $h(t)$ only approximates the characteristics of an ideal rectangular filter. Nevertheless, the band-pass character of the spectrum with a rather sharp roll-off close to the frequencies of 1 cps and 3 cps is clear from the lower part of Fig. 4.2. Outside the

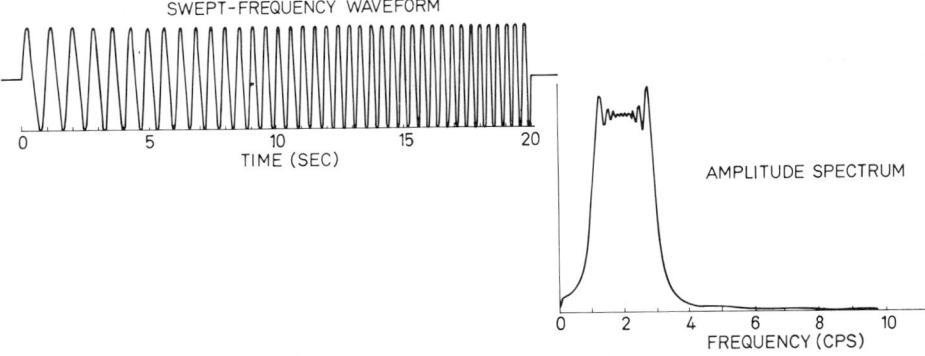

Fig. 4.2. Time-reversed unit-impulse response (upper part) and corresponding amplitude response (lower part) of a chirp filter. The frequency of the impulse response increases linearly from 1 cps to 3 cps in the time interval 0–20 sec.

sweep limits the amplitude spectrum decreases rapidly and may be considered as approximately equal to zero. Hence, cross-correlating a trace against a chirp waveform is equivalent to band-pass filtering. In spite of these excellent amplitude characteristics chirp filtering can introduce large phase distortion which can be eliminated by techniques discussed in Section 2.6. The sharpness of the roll-off may be increased without limits by taking longer and longer samples of the chirp waveform. Note, that the amplitude of $h(t)$ has been kept constant throughout the time interval considered. It would not be difficult to add different weights to different parts of the sample, thus further enhancing or suppressing particular frequencies in the pass band. However, theoretical conclusions applied to observational data (see Capon et al., 1969) show that the inclusion of amplitude variation has usually very little effect upon the signal/noise ratio improvement.

It follows from the foregoing discussion that for successful extraction of a weak signal from background noise by means of cross-correlation, some essential information concerning the signal and noise must be available a priori. The frequency composition and approximate arrival times of signals are especially important. Generally, the better our knowledge of the buried signal is, the better is the extraction that may be expected. Filters based upon a priori known or assumed statistical properties of the signal or noise are sometimes called *parametric detectors* (Cochran, 1973). When processing seismic records, frequency composition may be studied by analyzing less noisy traces recorded from similar sources or by nearby stations. Uniform instrumentation and homogeneous bedrock are naturally assumed. Recognition of multiple reflections can be based on an assumption of high similarity between the shapes of the primary and reflected pulses. Sometimes good resemblance can be observed between the recorded P- and pP-phases (Howell et al., 1967). Making use of this phenomenon, weak pP-onsets masked by the P-wave coda may be recovered from noisy records by cross-correlating the trace with the moving primary pulse. This technique operates equally well when pP shows opposite phase to P. Jones and Morrison (1954) demonstrated the effectiveness of the correlation-detection technique by processing an artificial record. They mixed a pilot pulse form with a recorded noise trace. The cross-correlation of the pulse against the mixture makes the pilot pulse clearly visible even on records with low signal/noise ratio.

4.2 MATCHED FILTERS

Next, we briefly discuss an alternative approach to correlation filters. In doing this, let us consider a linear system defined by its unit-impulse response $h_1(t)$ and let $x(t)$ and $y(t)$ be the input and output of the system, respectively. For any form of the input function we may define a system whose impulse response is the time-reversed replica of the applied input. Hence:

$$h_1(t) = x(-t) \qquad [4.6]$$

If $x(t)$ has a transient character with a gradual decay, the impulse response [4.6] obviously defines a noncausal system. Nevertheless, it is not difficult to modify $h_1(t)$ to make it causal. Due to its transient character, the input signal may be truncated at a certain time instant, t_1, beyond which the amplitudes of $x(t)$ may be neglected. The causal impulse response can then be written as:

$$h_2(t) = \begin{cases} x(t_1 - t) & \text{for } 0 \leq t \leq t_1 \\ 0 & \text{otherwise} \end{cases} \qquad [4.7]$$

Whereas [4.7] is of importance for physically realizable filters, for computer simulated systems the arbitrary delay term, t_1, can be ignored and the simpler relation [4.6] can be used.

Applying the convolution integral we determine the output signals at $t = t_0$, as:

$$y(t_0) = \int_0^\infty x(\tau) \, h_1(t_0 - \tau) \, d\tau = \int_0^\infty x(\tau) \, x(\tau - t_0) \, d\tau = R_{xx}(t_0) \qquad [4.8]$$

Relation [4.8] shows that the output is equal to the respective input autocorrelation function. In fact, this interesting phenomenon could have been intuitively expected since convolution and correlation differ from each other just in the time folding included in the former but absent in the latter. For real and even $h(t)$, which frequently has been the case in the present book, there is no difference between the two mathematical operations. Filters described by [4.6] and/or [4.7] are called *matched filters*. As follows from both equations, the impulse response of a matched filter related to a particular signal is a replica of the signal reversed in time. A tutorial survey of matched filters has been presented by Turin (1960).

Let us denote the Fourier transforms of the input and impulse response as $X(\omega)$ and $H(\omega)$, respectively. Making use of condition [4.6] and provided that $x(t)$ is a real function of time we get:

$$H_1(\omega) = \int_{-\infty}^\infty x(-t) \, e^{-j\omega t} \, dt = X^*(\omega) \qquad [4.9]$$

Ignoring the delay constant, introduced in [4.7] merely to make the filter physically realizable, the frequency response function of a matched filter, as expressed by [4.9], is the complex conjugate of the input spectrum. Therefore a matched filter is sometimes called a *conjugate filter*.

Matched filters represent an important category of filters due to the fact that they process signals in an optimum manner as far as the signal/noise ratio concerns. We shall briefly discuss this property further.

Different authors define the signal/noise ratio in different ways. Here, we shall adopt the definition that it is the ratio of the square of the signal amplitude at some time t_0, $|y(t_0)|^2$, to the mean-square amplitude of the output noise, $\overline{n_0^2}$. Thus, the signal/noise ratio becomes:

$$|y(t_0)|^2 / \overline{n_0^2} \qquad [4.10]$$

A filter which provides the highest signal/noise ratio is considered to be an optimum filter in a sense of this criterion. In order to determine the corresponding optimum frequency response function, defined as the Fourier transform of the impulse response, we determine the quantities in the numerator and denominator of the ratio [4.10]. In general, the output signal is given as:

$$y(t) = \mathcal{F}^{-1}\{Y(\omega)\} = \frac{1}{2\pi} \int_{-\infty}^\infty X(\omega) \, H(\omega) \, e^{j\omega t} \, d\omega \qquad [4.11]$$

where $X(\omega)$ and $H(\omega)$ are the input spectrum and response function of the filter, respectively. Making use of [2.46] and utilizing the input–output power spectrum relation [1.23], the quantity $\overline{n_0^2}$ may be expressed as:

$$\overline{n_0^2} = R_{oo}(0) = \frac{1}{2\pi} \int_{-\infty}^{\infty} \Phi_{oo}(\omega)\,d\omega = \frac{1}{2\pi} \int_{-\infty}^{\infty} H(\omega)\,H^*(\omega)\,\Phi_{ii}(\omega)\,d\omega \qquad [4.12]$$

where $R_{oo}(0)$ is the autocorrelation function of the output noise for zero delay, $\Phi_{oo}(\omega)$ is the output noise power spectrum and $\Phi_{ii}(\omega)$ is the input noise power spectrum.

Introducing [4.11] and [4.12] into [4.10] we write the signal/noise ratio in terms of input spectra $X(\omega)$ and $\Phi_{ii}(\omega)$ and the frequency response function $H(\omega)$. To evaluate the optimum $H(\omega)$ the ratio

$$\frac{|y(t_0)|^2}{\overline{n_0^2}} = \frac{\left| \int_{-\infty}^{\infty} X(\omega)\,H(\omega)\,\exp(j\omega t_0)\,d\omega \right|^2}{2\pi \int_{-\infty}^{\infty} H(\omega)\,H^*(\omega)\,\Phi_{ii}(\omega)\,d\omega} \qquad [4.13]$$

has to be maximized with respect to $H(\omega)$. The symbol t_0 in [4.13] denotes the time when $y(t)$ reaches its maximum amplitude. Rearranging the numerator in [4.13] and applying the Schwartz inequality (Turin, 1960; Tolstoy, 1973, p. 384) gives the optimum response function in a form:

$$H_{\mathrm{opt}}(\omega) = \frac{X^*(\omega)}{\Phi_{ii}(\omega)} \exp(-j\omega t_0) \qquad [4.14]$$

$H_{\mathrm{opt}}(\omega)$ maximizes the ratio [4.13]. If the input noise, $n_i(t)$, has a flat spectrum within the frequency interval occupied by the signal (white noise), [4.14] may be further simplified so that $H_{\mathrm{opt}}(\omega)$ becomes:

$$H_{\mathrm{opt}}(\omega) = KX^*(\omega)\exp(-j\omega t_0) \qquad [4.15]$$

where K is a positive constant defining the power spectrum level of the input noise. Factor K is of no importance in the design and may be neglected. Comparison of [4.15] and [4.9] shows that filters matched to their input signals perform an optimum processing of these signals in the sense of criterion [4.10]. The exponential factor in [4.14] and [4.15] reflects the time delay, t_0, in the corresponding impulse response. Matched filters specified by [4.9] give maximum signal/noise ratio only in the special case of white input noise. Usage of matched filters for seismic-reflection recording, assuming flat noise spectrum, has been described e.g. by Muir and Hales (1955). For nonwhite or coloured input noise, the filter has to be matched according to the ratio [4.14] to ensure the optimum processing. Recursive realization of matched filters has been outlined by Nikitin and Yanovskiy (1973).

It is obvious that these filters require an a-priori knowledge of the signal, $x(t)$, or its spectrum, $X(\omega)$. In practice, however, these quantities can usually be only estimated and thus the filters realized are merely approximations of true matched filters (see e.g. Yanovskiy, 1967, 1968; Shtemenko, 1971; Kats, 1972). In an ideal case, [4.8] shows that the output of a matched filter is the autocorrelation $R_{xx}(\tau)$, which usually has a form of a pulse consisting of several cycles with rather rapidly decreasing amplitudes. Though $R_{xn}(\tau)$ does not resemble the shape of the input signal it is convenient enough to reveal the existence of weak signals in a noise background. Treitel and Robinson (1969) described so-called *output-energy filters*. These filters do not consider only one particular output value, at $t = t_0$, as matched filters do, but maximize the sum of square-output values from a certain time interval.

Geophysical research provides a number of applications of matched filters. For example, Clay and Liang (1962) describe the adoption of filters matched to the source signal in a seismic-profiling system applied to deep-sea research.

Bruland and Rygg (1971) employed matched filters in the dispersion analysis of seismic surface waves. They assumed that linearly frequency-swept waveforms, discussed in the preceding section, match recorded surface waves sufficiently. Results presented by Capon and Green (1968) confirm that this assumption is quite reasonable. Chirp waveforms can therefore be efficiently used as impulse responses of corresponding matched filters. It is evident that in this case the processing performed may be considered either as matched or as chirp filtering. As follows from Fig. 4.3, matched filtering is a powerful tool for detecting arrivals of surface waves

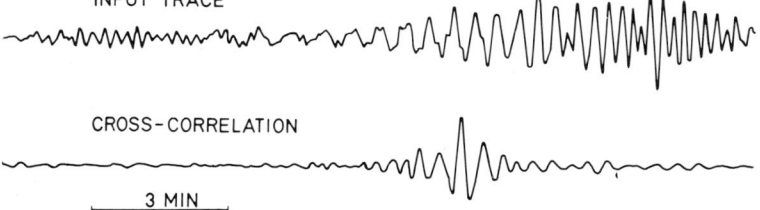

Fig. 4.3. Portions of recorded Rayleigh waves (upper trace) and the output of a matched (chirp) filter (lower trace). Maximum output amplitude indicates the arrival of a wave with 30-sec period. The duration of the chirp waveform used is about 630 sec and its frequency varies from about 0 02 cps to about 0 06 cps. After Bruland and Rygg (1971).

with prescribed periods. The performance of the filter with the spectacular result depicted in the figure is based upon an adequate resemblance between the chirp waveform and the analyzed signal. When this basic assumption is not satisfied the results obtained may easily lead to erroneous conclusions.

4.3 WIENER OPTIMUM FILTERS

For observational data where the signal and noise components occupy different

WIENER OPTIMUM FILTERS

frequency bands, the ideal rectangular filters, described in the preceding chapter, perform a perfect signal–noise separation. In practice, however, the signal and noise often possess overlapping spectra. Assuming overlapping spectra, a rectangular filter either cuts off a certain part of the useful information or passes through a certain part of the unwanted noise or, more likely, combines these two effects. Thus, for overlapping spectra, rectangular filters do not completely separate the signal from disturbing noise. Naturally, a question arises as to what then is the filter that performs the best signal–noise separation. In other words, how should the filter characteristics be chosen in order to suppress most of the noise and at the same time preserve most of the information carried by the signal? An exhaustive answer to this question is provided by the optimum-filter theory by Wiener (1949). Here, we briefly outline the approach as presented by Lee (1960) or Solodovnikov (1960).

4.3.1 *Optimum noncausal systems*

The problem to be solved may be defined in the following way: given the statistical characteristics of the signal and noise, design a filter which yields the best reproduction of the signal in agreement with a chosen criterion. In this section we assume the signal and noise to be stationary, noncorrelated, random functions of time. Further, we assume that the filter to be synthesized is linear and time-independent.

Consider the recorded trace, $x(t)$, as the sum of the signal and noise, so that:

$$x(t) = s(t) + n(t)$$

Due to the presence of noise, $n(t)$, a perfect recovery of the signal, $s(t)$, from $x(t)$ by means of a linear system is not possible. The *true (actual) output*, $y(t)$, of the system will always differ by some amount from $s(t)$. We assume that trivial cases, when $s(t)$ or $n(t)$ are known exactly, are excluded here. However, for the purpose of a theoretical derivation, we may introduce a quantity, $y_d(t)$, which equals the *ideal output* or *desired output*. How $y_d(t)$ will be defined depends upon our requirements. If we demand on the output an immediate copy of $s(t)$ then:

$$y_d(t) = s(t) \quad [4.16]$$

In various practical cases a pure time shift, by a constant, t_1, is not disturbing. Then:

$$y_d(t) = s(t \pm t_1) \quad [4.17]$$

where t_1 is a real nonnegative constant. Usage of a positive or negative sign in [4.17] leads to *prediction filters* or *lag filters*, respectively. In the former case we desire the advanced signal, $s(t + t_1)$, whereas in the latter case we desire the delayed signal, $s(t - t_1)$, to appear at the output of the synthesized system.

A prediction filter predicts future values of the signal provided that the histories of both the signal and noise are known. The system is designed so that when we feed the noisy trace $x(t)$ into the filter, its actual output will be the best approximation to the time-advanced signal $s(t + t_1)$. In a special case, $t_1 = 0$, see [4.16], we work with so-called *zero-lag filter* or *smoothing (enhancement) filter*.

There are a number of possible performance criteria for the designed filter. As follows from the discussion above, the desired output differs from the true output so that the instantaneous error, $e(t)$, is:

$$e(t) = y(t) - y_d(t) \qquad [4.18]$$

It is evident that any positive function of $e(t)$ may serve as a measure of how well the system operates. As examples, two reasonable choices, $|e(t)|$ and $\overline{e^2(t)}$ can be mentioned. The mean-square error criterion, used in the original work of Wiener (1949), weights large errors heavily while ignoring smaller ones but, on the other hand, it offers some mathematical advantages. The quantity $\overline{e^2(t)}$ can be determined in terms of the unit-impulse response of the filter and of the statistical characteristics of the input and output waveforms. Detailed investigation of $\overline{e^2(t)}$ in terms of the design parameters has been carried out by Wiggins (1967).

Substituting [4.17] into [4.18] and using the convolution integral to express the output signal, $y(t)$, the mean-square error becomes:

$$\overline{e^2(t)} = \lim_{T_1 \to \infty} \frac{1}{2T_1} \int_{-T_1}^{T_1} [h(\tau) x(t - \tau) \, d\tau - s(t \pm t_1)]^2 \, dt$$

where $h(t)$ is the unit-impulse response of the filter. Introducing correlation functions, the quantity $\overline{e^2(t)}$ can be written (see e.g. Lee, 1960, p. 359) as:

$$\overline{e^2(t)} = \int_{-\infty}^{\infty} h(\tau) \, d\tau \int_{-\infty}^{\infty} h(\sigma) \, d\sigma \, R_{xx}(\tau - \sigma) - 2 \int_{-\infty}^{\infty} h(\tau) \, d\tau \, R_{xs}(\tau \pm \tau_1) + R_{ss}(0) \qquad [4.19]$$

where $R_{xx}(\tau)$ is the autocorrelation of the filter input, $R_{xs}(\tau)$ is the input-desired-output cross-correlation and $R_{ss}(0)$ is the mean-square value of the desired filter output. Equation [4.19] documents an interesting phenomenon, namely that for any system $h(t)$ the mean-square error $\overline{e^2(t)}$ is completely determined by the correlation functions. The input and output waveforms themselves do not enter in [4.19]. Consequently, there are infinitely many input and desired-output waveforms which lead to the same mean-square error.

The next step is to optimize the system $h(t)$. This means to determine the impulse response which would minimize the expression [4.19]. This rather complicated mathematical problem is a subject for the calculus of variations. It can be shown, see e.g. Davenport and Root (1958, pp. 223–224), Lee (1960, pp. 360–369) or

WIENER OPTIMUM FILTERS

Solodovnikov (1960, pp. 186–188), that $h(t)$ which assures a minimum value for $\overline{e^2(t)}$ must satisfy the integral equation:

$$R_{xs}(\tau \pm t_1) - \int_{-\infty}^{\infty} h_{opt}(\sigma) R_{xx}(\tau - \sigma)\, d\sigma = 0 \qquad \text{for } \tau \geq 0 \qquad [4.20]$$

This equation, known as the *Wiener-Hopf equation*, provides conditions for determining the optimum mean-square error filter. To emphasize this point, we introduced the subscript 'opt' in the integrand in [4.20]. The limitation of [4.20] to the interval $\tau \geq 0$ and the lower limit of the integral follow from the requirement of causality and represent the basic difficulty in solving the Wiener-Hopf equation.

Fortunately, as in many cases before, for computer-simulated filters we may remove the condition of causality. Equation [4.20] is then replaced by:

$$R_{xs}(\tau \pm t_1) - \int_{-\infty}^{\infty} h_{opt}(\sigma) R_{xx}(\tau - \sigma)\, d\sigma = 0 \qquad \text{for all } \tau \qquad [4.21]$$

Since the second left-hand term in [4.21] is a convolution integral, the solution may be easily achieved by taking Fourier transforms of both sides of [4.21] which gives:

$$\Phi_{xs}(\omega) \exp(\pm j\omega t_1) - H_{opt}(\omega)\, \Phi_{xx}(\omega) = 0$$

where $H_{opt}(\omega)$ is the optimum system function and $\Phi_{xs}(\omega)$ and $\Phi_{xx}(\omega)$ are power spectra related to $R_{xs}(\tau)$ and $R_{xx}(\tau)$, respectively. The exponent t_1 makes the distinction between optimum lag ($t_1 < 0$), smoothing ($t_1 = 0$) and prediction ($t_1 > 0$) filters. The characteristics of the optimum filter are then:

$$H_{opt}(\omega) = \frac{\Phi_{xs}(\omega)}{\Phi_{xx}(\omega)} \exp(\pm j\omega t_1)$$

and:

$$h_{opt}(t) = \frac{1}{2\pi} \int_{-\infty}^{\infty} \frac{\Phi_{xs}(\omega)}{\Phi_{xx}(\omega)} \exp[j\omega(t \pm t_1)]\, d\omega \qquad [4.22]$$

Keeping in mind that the signal and noise are uncorrelated and assuming that either the signal or noise (or both) have zero mean [4.22] can be further simplified and become:

$$H_{opt}(\omega) = \frac{\Phi_{ss}(\omega)}{\Phi_{ss}(\omega) + \Phi_{nn}(\omega)} \exp(\pm j\omega t_1)$$

and:

$$h_{opt}(t) = \frac{1}{2\pi} \int_{-\infty}^{\infty} \frac{\Phi_{ss}(\omega) \exp[j\omega(t \pm t_1)]}{\Phi_{ss}(\omega) + \Phi_{nn}(\omega)}\, d\omega \qquad [4.23]$$

Since $\Phi_{ss}(\omega)$ and $\Phi_{nn}(\omega)$ are real functions of frequency, it is obvious that $H_{opt}(\omega)$ represents a phase-distortionless filter.

The main conclusion from our discussion can be formulated as follows: [4.23] define the Wiener optimum filter, assuming: (1) a linear, time-invariant system; (2) the signal and noise to be stationary random functions of time; (3) the signal and noise to be uncorrelated with zero means; (4) the minimum mean-square error criterion; and (5) a noncausal system.

The results outlined in this section have dealt only with noncausal systems. Physically-realizable optimum filters may be defined by solving the Wiener-Hopf equation [4.20]. This is an interesting but mathematically rather complicated procedure extending beyond the scope of the present book. The reader who might become interested in this and related problems is referred e.g. to an excellent presentation given by Lee (1960, pp. 371-395).

4.3.2 *Considerations in designing Wiener optimum filters*

From the point of view of practical applications, some of the properties of [4.21] and/or [4.23] have to be emphasized.

Firstly, we are able to determine experimentally only functions $\Phi_{xx}(\omega)$ and $\Phi_{nn}(\omega)$. The former can be evaluated as the power spectrum of the recorded trace, $x(t)$, the latter as the power spectrum of the noise preceding the signal arrival. Since observational data do not provide noiseless records, the power spectrum $\Phi_{ss}(\omega)$ can only be estimated. Due to this fact the synthesized filters only approximate the Wiener optimum filter.

Secondly, in practice it is rather an exceptional case that the signal and noise do not change their statistical characteristics with time. For example, when processing seismic records, the microseismic noise may be considered to some extent as stationary in time. Nevertheless, the signals recorded from earthquakes have evidently a transient, nonstationary character. Therefore, a filter designed for enhancement of a particular seismic signal will most likely provide rather poor results when applied to another seismic record. Generally speaking, the stationarity restriction, as formulated among the assumptions listed above, may be removed by introducing time-dependent, i.e. two-dimensional, power spectra in [4.23]. These spectra are then functions of frequency and time, thus depending upon absolute time as well as relative delay of signals or noise. The resulting filters then represent time-varying systems (see also Wang, 1969; Robinson, 1972). Instead of using the classical method of describing a system by its unit-impulse response, the system behaviour may be expressed as well by a set of linear differential equations. Whereas the former method has certain advantages when working with the so-called "black-box" representation, the latter becomes suitable for systems with an a priori known internal structure. The approach via a set of differential equations is used in the design of so-called *Kalman filters* (Bayless and Brigham, 1970; Ott and Meder,

1972). These filters are optimal in a Wiener sense and may be constructed without requiring the stationarity property.

Thirdly, so far we have considered analog signals and noise and consequently [4.23] define an optimum analog system. For digital computer processing, the results presented here have to be converted into a corresponding digital form. In order to be able to apply some of the methods explained in Chapter 2, function $H_{opt}(\omega)$ or $h_{opt}(t)$ has to be available in analytical form. However, in practice, functions $\Phi_{ss}(\omega)$ and $\Phi_{nn}(\omega)$ are usually estimated and measured, respectively, and their exact analytical expressions are unknown. It is therefore necessary to approximate $H_{opt}(\omega)$ by a proper analytical function prior to the analog–digital conversion (see e.g. Kulhánek, 1967). Several suitable approximation methods are described e.g. by Solodovnikov (1960, pp. 169–182). It is obvious that any approximation of $H_{opt}(\omega)$ results again in a concession from the optimum behaviour of the filter.

Figure 4·4 shows an example of an optimum smoothing filtering applied to a seismic record. The signal and noise on the original seismogram (upper trace) are of about the same level which makes an accurate visual detection of the signal onset difficult. After passing the trace through a filter which reasonably approximates the corresponding optimum system, the signal/noise ratio is significantly improved. The filter output (bottom trace) enables a fairly accurate reading of the first arrival. The filter used (for details see Kulhánek, 1967) is of the tenth order and introduces phase distortion. The sampling period is 0·138 sec and the analog–digital conversion has been carried out via the partial-fraction expansion.

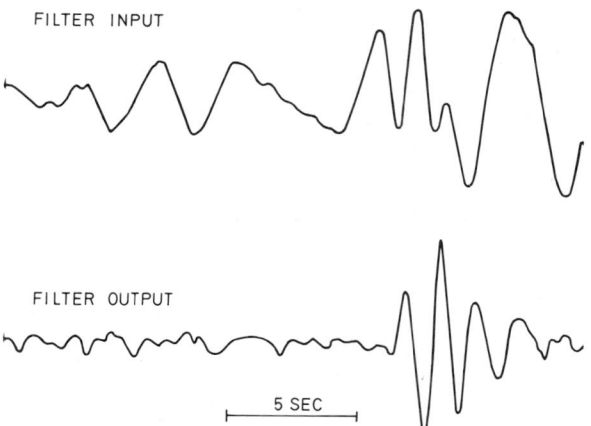

Fig. 4.4. Portions of a seismogram on the input (upper trace) and output (lower trace) of an optimum smoothing filter. The filter used introduces phase distortion.

It is also possible to carry out the entire synthesis directly in the discrete-time domain. Levinson (1949) makes use of the discrete form of the Wiener-Hopf equation (see also Robinson, 1967b; Shanks, 1967; Robinson and Treitel, 1967):

$$R_{xs}(m) - \sum_{n=-\infty}^{\infty} h_{\text{opt},n} R_{xx}(m-n) = 0 \quad \text{for } -\infty \leq m \leq \infty \quad [4.24]$$

where m and n are integer numbers. $R_{xs}(m)$ and $R_{xx}(m)$ are correlation functions defined as:

$$R_{xs}(m) = \lim_{N \to \infty} \frac{1}{2N+1} \sum_{n=-N}^{N} x_n s_{n+m}$$

and:

$$R_{xx}(m) = \lim_{N \to \infty} \frac{1}{2N+1} \sum_{n=-N}^{N} x_n x_{n+m}$$

Assuming that $h_{\text{opt},n}$ has a transient character, then for suitably chosen N [4.24] may be modified as:

$$R_{xs}(m) - \sum_{n=-N}^{N} h_{\text{opt},n} R_{xx}(m-n) = 0 \quad \text{for } -N \leq m \leq N \quad [4.25]$$

Hence, the condition for filter optimization is expressed by means of a set of $2N+1$ equations performing the same function as the Wiener-Hopf integral equation does in the continuous-time domain. Coefficients of these equations are the known values of the input autocorrelation, input–desired-output cross-correlation and unknown are the optimum-impulse-response values. The solution determines the optimum sequence $\{h_{\text{opt}}\}$ of a length $2N+1$. Thus, the solution of the integral for analog systems has been substituted in the discrete-time domain by the solution of a system of linear algebraic equations. An approximate solution of [4.25] by means of iteration techniques has been described by Wang and Treitel (1973).

Application of the z-transform to [4.24] yields:

$$R_{xs}(z) = H_{\text{opt}}(z) R_{xx}(z)$$

so that the pulse-transfer function of the optimum filter becomes:

$$H_{\text{opt}}(z) = \frac{R_{xs}(z)}{R_{xx}(z)} = \frac{R_{ss}(z)}{R_{ss}(z) + R_{nn}(z)} \quad [4.26]$$

where $H_{\text{opt}}(z)$, $R_{xs}(z)$ and $R_{xx}(z)$ are the z-transforms of corresponding functions entering [4.24]. Equation [4.26] provides the optimum filter characteristics directly in the digital form. Note, that $R_{xs}(z)$ and $R_{xx}(z)$ in [4.26] are both in general infinite-length series and therefore in practice the truncation problem necessarily enters the design. Furthermore, it is usually easier to make certain assumptions about the signal and noise forms in the frequency domain, [4.23], rather than in the time domain as required in [4.26].

In addition to the approach used by Wiener (1949) and Levinson (1949) a number of other methods have been employed to construct optimum smoothing and prediction operators. For example, Frank and Doty (1953) revealed that the optimum signal/noise ratio improvement is obtained by a filter whose amplitude response has a peak at the frequency corresponding to the maximum signal/noise ratio.

The Geophysical Analysis Group at the Massachusetts Institute of Technology (Wadsworth et al., 1953) used a linear-prediction operator to detect hidden reflections on seismic records. The idea consists basically in an assumption that stationary portions (nonreflection intervals) of traces are predictable with significantly higher accuracy when compared with portions during which the reflections occur. In other words, time intervals with large prediction errors coincide with reflection arrivals. The error of prediction is defined as the amplitude difference between the predicted and true trace. The dependence between the prediction error and the length of the prediction operator is studied by Galbraith (1971). Wadsworth et al. (1953) found that prediction operators based upon past, stationary, portions of records from two adjacent geophones are superior to operators constructed from a single trace. Consequently, operators applied may be considered as multiple-input prediction operators. In several numerical examples, which they present, the sampling period used is $\Delta t = 2 \cdot 5$ msec. Coefficients of the optimum operator are determined from 125 msec long trace samples as a solution of a set of linear algebraic equations. Entries of these equations are the known history of the two input traces together with unknown operator coefficients. Prediction times are 5, 10 and 15 msec and the advanced output values are predicted from present and the three most recent values of both input traces. A brief explanation of the reported technique could be also found in Swartz and Sokoloff (1954). A similar concept has been employed by Moltshan et al. (1964) in a single-channel detection of weak seismic signals and by Claerbout (1964) in a multi-channel processing.

Robinson (1957, 1967a) constructed a single-input prediction operator. He introduced an assumption that the processed trace is composed of a number of wavelets of the same shape but with different amplitudes and arrival times. Robinson calls the approach the *predictive decomposition*.

Simpson (1955) developed a linear optimum operator based upon an assumption that signals on multiple-trace seismograms are identical in shape and no time delays take place. Noise is considered to be represented by a stationary time function. The optimization of the filter is established by a transformation which makes the filtered traces most "similar" at least for time intervals when the signals occur.

An optimum prediction operator to forecast the occurrence of stronger rockbursts in a limited mining area has been developed by Buben and Rudajev (1974). They assume a random character and mutual independence of time intervals between and energies of individual rockbursts. The original time sequence of measured seismic activities, in this case a number of shocks within a certain energy interval per month, was smoothed by applying the moving averaging. The smoothed sequence yields a linearly decreasing autocorrelation function defining the coefficients of the prediction operator.

4.3.3. *Multi-dimensional Wiener optimum filters*

The theory of Wiener optimum filters may be extended from one-dimensional to multidimensional filters. In geophysical applications especially, two- and three-dimensional optimum filters are of great importance. The former are usually vector wavenumber or wavenumber and frequency filters, whereas the latter are vector wavenumber and frequency filters. The optimum two-dimensional transfer function is given by:

$$H_{opt}(k_x, k_y) = H_{opt}(\mathbf{k}) = \Phi_{xs}(k_x, k_y)/\Phi_{xx}(k_x, k_y) \qquad [4.27]$$

where k_x and k_y are e.g. wavenumbers in the x- and y-direction respectively, $\mathbf{k}$ is the vector wavenumber, $\Phi_{xs}(k_x, k_y)$ is the input-desired-output cross-power spectrum and $\Phi_{xx}(k_x, k_y)$ is the input power spectrum. The power spectra may be expressed in terms of corresponding two-dimensional correlation functions. The impulse response function related to [4.27] is again a two-dimensional function defined over the (x,y) plane. Clarke (1969) applied the filter [4.27] to improve gravity and magnetic maps. The filters Clarke derived perform a Wiener sense optimum downward continuation and second differentiation of mapped data. Except for the reference mentioned, the theory of two-dimensional optimum filters has also been discussed in different connections (see e.g. Chapter 6), by Wiggins (1966) and Sengbush and Foster (1968) among others. A time-domain approach and use of a two-dimensional Hamming window has been outlined by Gunn (1972).

Extending further, the transfer function of an optimum three-dimensional filter becomes:

$$H_{opt}(f, k_x, k_y) = H_{opt}(f, \mathbf{k}) = \Phi_{xs}(f, k_x, k_y)/\Phi_{xx}(f, k_x, k_y) \qquad [4.28]$$

where $\Phi_{xs}(f, k_x, k_y)$ and $\Phi_{xx}(f, k_x, k_y)$ are the input–desired-output and input power spectra, respectively. These power spectra may be estimated via the three-dimensional Fourier transforms of corresponding correlation functions. The theory for three-dimensional Wiener filters has been adequately treated by Burg (1964). It is evident that [4.27] and [4.28] are generalizations of the one-dimensional optimum filter [4.25].

4.4 POLARIZATION FILTERS

As will be emphasized below, polarization filtering differs significantly from filtering procedures discussed in the preceding sections of the present book. However, *polarization filters* represent an interesting approach in making use of correlation techniques for signal enhancement and so a brief description has been included in this chapter.

Three-component seismograms permit treatment of the recorded seismic signals as time vectors. With records from a matched set of three seismometers on hand *particle-motion patterns* (*polarizations*) of arriving seismic waves can be constructed and analyzed. The particle motion related e.g. to propagating P-waves is, as commonly known, *rectilinearly polarized* in the direction of propagation. This direction, beneath the detector, depends upon the location of the focus and the recording-site structure.

From the filtering point of view, the polarization phenomenon may be utilized in two different ways. Firstly, it is possible to distinguish signals according to their different types of polarization, i.e. to separate rectilinearly polarized signals and signals with other, e.g. *elliptical*, *polarizations*. Secondly, recorded P-wave motions may be discriminated due to the direction of polarization. It is then possible to pick out one particular focal region from a number of regions, which may be used in constructing automatic alarm systems (see e.g. Levin and Price, 1964).

Generally, the efficiency of both types of filters depends upon the degree of *rectilinearity* and *directionality* of particle-motion patterns at the recording site (see e.g. Flinn, 1965; Montalbetti and Kanasewich, 1970). In this section we present principles of polarization filters used for separating recorded P-waves from strong background noise.

Consider that the signal of interest is a rectilinearly polarized P-wave. Let the nth digital value of the two horizontal and the vertical components be s_{1n}, s_{2n} and s_{3n}, respectively. The noise will be represented here by microseisms with components denoted as n_{1n}, n_{2n} and n_{3n}. Provided that microseisms propagate, in principle, as Rayleigh waves, the microseismic particle motion is elliptically polarized. The three traces become:

$$x_{1n} = s_{1n} + n_{1n}, \quad x_{2n} = s_{2n} + n_{2n} \quad \text{and} \quad x_{3n} = s_{3n} + n_{3n}$$

Due to the rectilinear polarization, the phase shift between the signal components is ideally equal to 0° or 180°. Consequently, cross-correlating any two of the three signal components yields normalized values 1 or -1. In general, no significant resemblance is expected between the noise components and therefore the absolute noise cross-correlations will be significantly less than 1. Let us cross-correlate e.g. the two horizontal traces. We have:

$$R_{12}(\tau) = R_{s_1 s_2}(\tau) + R_{s_1 n_2}(\tau) + R_{s_2 n_1}(\tau) + R_{n_1 n_2}(\tau) \qquad [4.29]$$

The first term on the right is equal to 1 or -1 whereas the remainder may be neglected, provided that there is no significant correlation between signal and noise. For more details see e.g. Shaub (1963).

Equation [4.29] describes the procedure performed by a polarization filter and shows how to detect rectilinearly polarized signals buried in the noisy, nonlinearly polarized background. As an input to the filter any two of the three components may be used. At the output large amplitude values are observed at or close to time instants corresponding to signal arrivals.

In general, cross-correlation functions $R_{12}(\tau)$, $R_{13}(\tau)$, $R_{23}(\tau)$ do not provide high time resolution, thus introducing certain inaccuracy in determining the signal arrivals. Furthermore, cross-correlation merely measures the rectilinearity of polarization and does not take into account the total power of signals involved. The time resolution is governed by the length of time windows applied. The shorter samples provide higher resolution whereas the longer samples provide better suppression of nonlinearly polarized trace components. Reasonable compromise has to be chosen in each particular case.

Several practical applications of polarization filters have been described by Shimshoni and Smith (1964). They utilize the horizontal radial, x_{4n}, and vertical, x_{3n}, components and instead of cross-correlation they make use of a cross-product M_j defined as:

$$M_j = \sum_{n=-N}^{N} x_{3n+j} \, x_{4n+j} \qquad [4.30]$$

where $2N + 1$ is the length of the time window used. The product M_j gives again a measure of the rectilinearity. Besides that, it also preserves the power of signals involved. In case of perfect rectilinearity M_j is a positive quantity (x_{3n} and x_{4n} are always in phase) which with respect to the time resolution and further analysis is not the most advantageous form. Shimshoni and Smith therefore tested several other functions of M_j, namely $x_{3j}M_j$, $x_{4j}M_j$, $(x_{4j}M_j)^{\frac{1}{2}}$ and $x_{4j}|M_j|^{1/2}$ with obviously superior results. They mixed signals recorded from a large Nevada underground nuclear explosion with noise samples of varying energy level. Thus producing traces with different signal/noise ratios, the efficiency of the filter has been tested.

Assuming the signal to be rectilinearly and the noise nonlinearly polarized, we have been considering an ideal record. In practice, however, this will rarely be the case. For example, signal-generated noise (reverberations at the recording site) and the crustal structure beneath the receiver will contribute to a departure from the perfect rectilinearity.

As mentioned above, polarization filters essentially differ in several respects from most of the filters discussed previously in this book. Firstly, the signal–noise discrimination is no more based upon different frequency contents but upon different types of polarization. Hence, also signals and noise occupying the same frequency band but being differently polarized can be effectively separated from each other. Secondly, since [4.29] and [4.30] do not satisfy the proportionality property [1.2], polarization filtering is a nonlinear procedure as was e.g. the case with autocorrelation filtering. Thirdly, there are always two input quantities (components) entering the polarization filtering process. Hence, polarization filters can be visualized as an example of simple multiple-input filters.

Mercado (1968) constructed a linear version of a polarization filter for a separation between P-waves and Rayleigh waves. Whereas the former appear with the same aligned shapes on both the vertical and horizontal channels the latter shows

90° delay between the channels. Time shifts, between the two components, due to the 90° delay are frequency-dependent and may be utilized in the mode separation. Two-channel filters are designed to pass signals with no time shift and to reject signals with specified time delays between the two channels. Choy and McCamy (1973) constructed a polarization filter to enhance records of signals polarized as Rayleigh waves.

Chapter 5

DECONVOLUTION FILTERS

Removing the undesired effects of sensor characteristics is common to many fields of science. In geophysics, e.g., seismic records provide time series which differ in general from the true ground motion. This becomes a rather serious problem especially when traces from different sensors are to be compared. Minimizing the sensor influence by using an additional system with inverse characteristics usually amplifies the noise which may thus come to dominate the record. One has to compromise between the noise amplification and undesired instrumental effect. Besides, exact inverses of nonrecursive systems are usually of infinite length which obviously makes computer processing impossible. Since most geophysical systems are of band-pass type, the exact inverse necessarily provides an unstable operator, irrespective of whether the original system is of recursive or nonrecursive type. This is another major difficulty in constructing inverse systems. In some applications the recorded traces may be treated as a superposition of similar time-delayed wavelets. If this is the case, the record resolution may be improved by decomposing the trace into its original wavelets, using a deconvolution filter.

This chapter treats practical realizations of deconvolution of nonrecursive filters separately for minimizing the sensor effect and for record-resolution improvement.

5.1 EXACT DECONVOLUTION

Consider a linear time-invariant system defined by its impulse response $h = (h_0, h_1, \ldots)$, which in general will be of infinite length. If x and y are the input and output sequences, respectively, then the three series are related in the time domain via the convolution summation given in [1.41]. In the frequency domain, [1.45], we have the corresponding relation, $Y(z) = H(z) / X(z)$. A *deconvolution filter* or *inverse filter* may be represented by a system with characteristics $D(z)$ and d which, when excited by the signal y, will respond with the original signal x (see Fig. 5.1). If

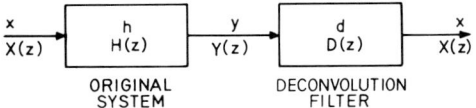

Fig. 5.1. Block-diagram representation of exact deconvolution (inverse) filtering.

EXACT DECONVOLUTION

this is the case, then the inverse filter in Fig. 5.1 completely removes the effect of the system h on the input signal. The problem to be solved is the design of $D(z)$ or d.

Several basic relations follow immediately from the block diagram in Fig. 5.1. Applying the input–output formulae to the original system we have:

$$y = x * h \quad \text{and} \quad Y(z) = X(z) H(z) \tag{5.1}$$

As usual the symbol $*$ is used to designate the convolution operation. The inverse filter yields:

$$x = y * d \quad \text{and} \quad X(z) = Y(z) D(z) \tag{5.2}$$

Combining [5.1] and [5.2] we obtain:

$$x = (x * h) * d \quad \text{and} \quad X(z) = X(z) H(z) D(z)$$

From the last equation we gain essential relations between the original and related inverse system, namely:

$$H(z) D(z) = 1 \tag{5.3}$$

$$h * d = \delta \tag{5.4}$$

where δ is the unit-impulse sequence, $(1, 0, 0, \ldots)$, discussed in Section 1.6. Let the impulse-response sequence of the original system, normalized to give $h_0 = 1$, be $h = (1, h_1, h_2, \ldots)$. Introducing h into [5.4] we write:

$$(1, h_1, h_2, \ldots) * (d_0, d_1, d_2, \ldots) = (1, 0, 0, \ldots)$$

Usage of any of the four convolution schemes, mentioned in Section 1.7 leads to:

$$(d_0, d_1 + d_0 h_1, d_2 + d_1 h_1 + d_0 h_2, \ldots) = (1, 0, 0, \ldots)$$

Comparing the samples corresponding to the same time instants on both sides of the equality yields a set of linear equations in the unknown inverse filter coefficients:

$$\left. \begin{array}{l} d_0 = 1 \\ d_1 + d_0 h_1 = 0 \\ d_2 + d_1 h_1 + d_0 h_2 = 0 \\ \quad \cdot \\ \quad \cdot \\ \quad \cdot \end{array} \right| \tag{5.5}$$

Substituting $d_0 = 1$ in the second equation we solve this for d_1. Substituting d_1 in the third equation we obtain d_2, etc. The impulse response we are looking for has a form $d = (1, -h_1, h_1^2 - h_2, \ldots)$. In general, [5.5] is a set of infinitely many equations for an infinitely long inverse-filter impulse response.

It follows immediately from [5.3] that polynomial long division leads to:

$$\begin{aligned} D(z) &= d_0 + d_1 z^{-1} + d_2 z^{-2} + \ldots = \frac{1}{1 + h_1 z^{-1} + h_2 z^{-2} + \ldots} \\ &= 1 - h_1 z^{-1} + (h_1^2 + h_2) z^{-2} - \ldots \end{aligned} \tag{5.6}$$

Equating coefficients of like powers of z results obviously again in the impulse response $d = (1, -h_1, h_1^2 - h_2, \ldots)$. Summarizing, the polynomial multiplication [5.3] as well as the convolution [5.4] provide the infinitely long impulse response of the exact inverse filter.

The next problem to be solved is twofold. Firstly, the response d due to its infinite length cannot be processed by means of digital computers. Secondly, without any additional restriction, we have no guarantee that d and $D(z)$ represent a stable system (see Section 1.8). Stability and finite length of inverse systems are of course necessary requirements.

5.2 TRUNCATED APPROXIMATE DECONVOLUTION

Methods for constructing approximate systems to obtain finite length and stable inverse filters are called for. Corrective filters may be added to exact inverses in order to keep the total gain constant (Berckhemer and Jacob, 1968) or zero (Bogert, 1962; Burch et al., 1964) for low and high frequencies. An iterative technique solving the problem in the time domain has been outlined by Brigham et al. (1968) and Neunhöfer (1971).

Another approach based upon the concept of minimum- and maximum-delay response as presented by Treitel and Robinson (1964) and Robinson (1967b) will be discussed in more detail.

We start the stability investigation by considering a simple system with a normalized impulse response made up of only two members, i.e. $h = (1, h_1)$. Robinson (1967b) calls such wavelets *normalized dipoles*. Let us investigate the behaviour of an inverse filter d, $D(z)$ corresponding to this dipole. Taking advantage of the result given by [5.6] and substituting $h_n = 0$ for $n = 2, 3, \ldots$, the impulse response becomes:

$$d = (1, -h_1, h_1^2, -h_1^3, \ldots) \quad [5.7]$$

and the corresponding system function may be written as:

$$D(z) = 1 - h_1 z^{-1} + h_1^2 z^{-2} - h_1^3 z^{-3} + \ldots$$

Equation [5.7] is of immediate use as far as the system stability is concerned. It is obvious that for $|h_1| < 1$ the system is stable, whereas for $|h_1| > 1$ it is unstable. According to Robinson and Treitel (1964), the response $h = (1, h_1)$ in the former case is called *minimum-delay response* and the corresponding inverse, d, is stable. In the latter case h is called *maximum-delay response* and the corresponding inverse response is unstable.

The stability of the inverse of a normalized dipole may be investigated in the frequency domain as well. As follows from [5.6] the system function here becomes $H(z) = 1 + h_1 z^{-1}$ and the corresponding inverse $D(z) = 1/(1 + h_1 z^{-1})$. Referring to Fig. 1.7 the coefficient $|h_1| < 1$ provides a stable inverse while $|h_1| > 1$ generates an unstable inverse filter.

There is a possibility to make the inverse of a maximum-delay response stable by employing an *anticipation function*, i.e. one that operates only on future input values. Consider a normalized dipole, $h = (1, h_1)$ where $|h_1| > 1$, and the impulse response, $d = (\ldots, d_{-3}, d_{-2}, d_{-1})$, of a deconvolution filter. Introducing h and d into [5.3] we obtain:

$$(\ldots, d_{-3}z^3 + d_{-2}z^2 + d_{-1}z) = 1/(1 + h_1 z^{-1})$$

Performing the long division, the right-hand side may be expressed in terms of positive powers of z so that:

$$(\ldots, d_{-3}z^3 + d_{-2}z^2 + d_{-1}z) = h_1^{-1}z - h_1^{-2}z^2 + h_1^{-3}z^3 - \ldots$$

Since we have assumed $|h_1| > 1$, the response of the deconvolution filter

$$d = (\ldots, h_1^{-5}, -h_1^{-4}, h_1^{-3}, -h_1^{-2}, h_1^{-1}) \qquad [5.8]$$

represents a stable system. Broadly speaking, it is possible to construct a stable inverse filter irrespective of whether the original filter $h = (1, h_1)$ is a minimum- or maximum-delay system. In the former case the solution is provided in terms of a causal system (impulse response is of a memory type) while in the latter case the impulse response we are looking for is of an anticipation type, describing a non-causal system.

The output value, x_n, of the inverse filter [5.7] is:

$$x_n = \sum_{m=0}^{\infty} d_m y_{n-m} \qquad [5.9]$$

and for the inverse filter [5.8], the output values become:

$$x_n = \sum_{m=-\infty}^{-1} d_m y_{n-m} \qquad [5.10]$$

In both cases, the deconvolution filters are stable systems and consequently the output sequences [5.9] and [5.10] converge. Thus, it is possible to obtain sufficiently accurate output values even with the two series truncated at a reasonable point, so-called *truncated approximate inverses*. After replacing the infinite limits in [5.9] and [5.10] by suitable finite constants, the concept of inverse filtering, outlined above, is ready to be processed by a digital computer.

In practice, however, an impulse response in the form of a normalized dipole represents a rather special case. Let us therefore investigate a more general case when the original system is defined by an impulse response, $h = (h_0, h_1, \ldots, h_M)$, of length $M + 1$. Corresponding system function is:

$$H(z) = h_0 + h_1 z^{-1} + \ldots + h_M z^{-M} \qquad [5.11]$$

The polynomial [5.11] may be expanded into a product of M dipoles and written in a form:

$$H(z) = (\alpha_0 + \alpha_1 z^{-1})(\beta_0 + \beta_1 z^{-1}) \ldots (\epsilon_0 + \epsilon_1 z^{-1}) \qquad [5.12]$$

where $\alpha_0, \alpha_1, \beta_0, \beta_1, \ldots, \epsilon_0, \epsilon_1$ are given by roots of the polynomial [5.11]. Both real and complex values may appear. As long as $h_0, h_1, \ldots, h_M$ are real constants, complex roots appear always in complex-conjugate pairs. Thus a real-valued input always generates a real-valued output. The factored system function [5.12] may be visualized as a serial arrangement of M first-order nonrecursive filters. Corresponding impulse response may be then expressed in terms of M successive convolutions as:

$$h = (\alpha_0, \alpha_1) * (\beta_0, \beta_1) * \ldots * (\epsilon_0, \epsilon_1) \qquad [5.13]$$

Each of the dipoles entering the cascade of M convolutions in [5.13] is a minimum- or maximum-delay wavelet. The appropriate inverse and stable wavelets have been defined in [5.7] and [5.8]. A successive convolution of these M inverses provides the impulse response d which is the stable inverse of the original response h. After reaching sufficiently small values, the stable inverse, d, may be truncated, thus making the computer processing possible.

If each of the dipoles, entering [5.13], is of minimum-delay type, then the composite h is a minimum-delay response. Similarly, if all dipoles are maximum-delay dipoles, then h is a maximum-delay response. A mixture of maximum- and minimum-delay dipoles generates a *mixed-delay response*. According to Robinson and Treitel [1965] a minimum-delay response has most of its energy concentrated at the beginning of the sequence representing the response. A maximum-delay response is the one with most of its energy concentrated at the end of the sequence. Mixed-delay responses are responses between the two extremes.

5.3 LEAST-SQUARES APPROXIMATE DECONVOLUTION

The simple truncation of the exact inverse response, mentioned in the preceding section, may provide a reasonable, but not the best approximation to the inverse filter. There are inverse filters of the same length generating output signals which approximate the desired form with less error than may be achieved by simply truncated inverses. Below, we investigate inverses of finite length that provide the best approximation, in least-squares sense, of the desired output. These filters are called *least-squares approximate inverse filters*.

As an example consider a simple case. Let the investigated inverse be a one-term response, $d = (d_0)$, and let the original system be described by a minimum-delay response $h = (1, h_1)$ where $|h_1| < 1$. The original system is excited by a unit impulse $x = (1, 0, \ldots)$. Since the convolution of a signal with the unit impulse is the signal itself, the original system will respond with $y = x * h = h$ (see Fig. 5.1). The original and inverse filter perform a serial arrangement so that at the output of the inverse filter we obtain:

$$g = y * d = h * d = (1, h_1) * d_0 = (d_0, d_0 h_1)$$

Due to the finite length of the inverse filter, the true output, g, and the desired output, x, differ from each other so that the error sequence becomes:

$$e = x - g = (1, 0, \ldots) - (d_0, d_0 h_1)$$
$$= (1 - d_0, -d_0 h_1) = (e_0, e_1) \qquad [5.14]$$

The next step will be the determination of the value d_0 which minimizes the error [5.14] according to the chosen criterion. We determine the error energy, as the sum of squares, $e_0^2 + e_1^2$, and equate its derivative, with respect to d_0, to zero:

$$\frac{\partial}{\partial d_0} \left[(1 - d_0)^2 + (-d_0 h_1)^2 \right] = 0$$

It follows from this equation and from the positive second derivative that the condition for the optimum error energy inverse may be expressed as:

$$d_0 = 1/(1 + h_1^2) \qquad [5.15]$$

Some authors (see e.g. Robinson, 1967b; Kanasewich, 1973) prefer to express the optimum inverse in terms of the autocorrelation, $R_{hh}(n)$, of $h = (1, h_1)$. The autocorrelation has three nonzero terms:

$$R_{hh}(n) = \sum_{k=-\infty}^{\infty} h_{k+n} h_k = (\ldots, 0, h_1, 1 + h_1^2, h_1, 0, \ldots) = (R_{-1}, R_0, R_1)$$
$$\uparrow$$

where the vertical arrow indicates the time origin $n = 0$ (see also Section 1.6). Introducing the value of autocorrelation for the argument $n = 0$ into [5.15] we have:

$$d_0 = 1/(1 + h_1^2) = 1/R_{hh}(0) = 1/R_0$$

Applying the least-squares approach to a two-term inverse, $d = (d_0, d_1)$, results in:

$$d_0 = \frac{1 + h_1^2}{1 + h_1^2 + h_1^4} = \frac{R_0}{R_0^2 - R_1^2}$$

$$d_1 = \frac{-h_1}{1 + h_1^2 + h_1^4} = \frac{-R_1}{R_0^2 - R_1^2} \qquad [5.16]$$

The same procedure may be extended to a case where the original impulse response is a minimum-delay response of arbitrary length, say, $h = (h_0, h_1, \ldots, h_M)$. Suppose that the exact inverse will be approximated, via the least-squares approach, by a response of the length $N + 1$, i.e. $d = (d_0, d_1, \ldots, d_N)$. The $N + 1$ unknowns may be determined from a set of so-called normal equations (Robinson, 1967b, p. 170):

$$d_0 R_0 + d_1 R_1 + d_2 R_2 + \ldots + d_N R_N = h_0$$
$$d_0 R_1 + d_1 R_0 + d_2 R_1 + \ldots + d_N R_{N-1} = 0$$
$$\vdots \qquad \vdots$$
$$d_0 R_N + d_1 R_{N-1} + d_2 R_{N-2} + \ldots + d_N R_0 = 0 \qquad [5.17]$$

where $R_0, R_1, \ldots, R_N$ are autocorrelation coefficients of the original impulse response h.

One important behaviour of the least-squares deconvolution outlined in this section should be accentuated. Note that [5.15], [5.16] and [5.17] determine optimum and stable inverses, provided that the original system has a minimum-delay response, h. These equations may be applied to maximum- and mixed-delay responses as well but the resulting inverses will not be optimal in the given sense. In the above discussion we desired the inverse filter to respond by $x = (1, 0, \ldots)$, i.e. by a unit impulse appearing at time index 0. It follows from experimental data, presented by Robinson (1967b, pp. 180-182), that in the case of a system with a maximum- or mixed-delay response, better approximate inverses are obtained for delayed output impulses, i.e. for $x = (0, 1, 0, \ldots)$, $x = (0, 0, 1, 0, \ldots)$, etc. The corresponding inverse filter is then called *delayed inverse filter* or *delayed spiking filter*.

For most geophysical applications a known time shift in the output signal is fully acceptable. When h is a maximum-delay wavelet, the least-error energy approximation is found by shifting the output impulse to the end of the output signal. Consider e.g. the original response as a maximum-delay two-term response and the desired inverse as a three-term response. The true output of the inverse filter contains $2 + 3 - 1 = 4$ terms. Then, the best approximation of the exact inverse is obtained by employing a spike at the time index 3, $x = (0, 0, 0, 1,)$, as the desired output. For mixed-delay systems the spike has to be placed at some intermediate position in order to perform the least-error energy deconvolution (see also Treitel and Robinson, 1966b; Kondrat'yev, 1968). Matrix-form solution of the problem is given and discussed by Rice (1962). Ford and Hearne (1966) present a solution of an inverse filter which transforms a minimum-delay wavelet defined by its autocorrelation function into a unit impulse. Deconvolution operators which are optimum with respect to errors in correlation functions and powerspectra estimates are described by Burns (1968). In order to sharpen the P- and pP-phase onsets, Frasier (1972) applied the deconvolution technique to Nevada explosions. Davies and Mercado (1968) extended the method of deconvolution filtering to multi-channel systems. However, when field data are processed, multi-channel deconvolution shows little improvement when compared with single-channel deconvolution. Principles of digital multiple-input–multiple-output inverse filtering have been well treated by Treitel (1970).

The least-squares approach discussed above may also be employed in cases where the desired output is no more an impulse (delayed or with zero delay) but an arbitrarily given response. The resulting inverse filter converts, in the least-error energy sense, h into a prescribed form and is therefore called *shaping filter*. Solution for a shaping filter may be found from [5.17] by substituting the right-hand sides by cross-correlation coefficients between the input h and desired output (Robinson, 1967b, p. 176). We remark that the estimation of these coefficients as well as those entering [5.17] is a complicated problem in itself. Difficulties are caused mostly by the finite length and nonstationarity of the recorded wavelets (Burns, 1968; Foster et al., 1968).

5.4 TRACE DECOMPOSITION

Besides the removal of instrumental effects, principles of digital deconvolution can be used to improve the record resolution (see e.g. d'Hoeraene, 1962; Rice, 1962; Kunetz and Fourmann, 1968). When analyzing seismic records, the overlapping seismic phases and their interference complicate measurements of later arrivals. If the duration of the individual phases on the seismogram could be adequately shortened, later arrivals would appear on the record in time when trace oscillations due to preceding seismic phases have already disappeared. Hence, the determination of onsets related to individual phase arrivals would be easier.

The essential assumption made in this section is that due to the same source mechanism, all seismic phases in, say, a P-wave group have similar and reasonably simple form. Since waves such as P, pP, PcP, PP or multiple reflections within surface layers, etc. arrive at the receiver via different paths, form changes between individual phases and, even more importantly, time delays between phase arrivals will occur on the record. The final record may be visualized as a superposition of these disturbed and time-delayed phases. It will be assumed that the noise effect is small and it will be neglected, hereafter.

Changes in the form and time delays of individual phases may be expressed by means of a system function, $H_i(z)$, or corresponding impulse response, h_i, where i is a seismic phase index. Let us assume u be a time sequence representing the original form common to all studied phases. The sequence u may be considered as a *primary* or *source wavelet*. The ith phase propagates along a path characterized by h_i or $H_i(z)$ so that this phase will appear on the seismogram as:

$$s_i = u * h_i \qquad i = 1, 2, \ldots, N$$

where the symbol ∗ denotes convolution and N is the total number of phases considered. The given signal, s, which is a sum of all individual phases, s_i, may be written as:

$$s = \sum_{i=1}^{N} s_i = \sum_{i=1}^{N} u * h_i \qquad [5.18]$$

Similar approach, applied to multiple-reflection records, has been suggested e.g. by Sakrison et al. (1967). They define the signal sequence s in terms of the reflection coefficients of different layers and time delays corresponding to reflections from these layers.

Due to the fact that convolution is distributive over addition, [5.18] may be expressed in a form:

$$s = u * \sum_{i=1}^{N} h_i = u * h \qquad [5.19]$$

where h is the total impulse response. For a given sequence u there is an explicit correspondence between s and h. Proper selection of u, sometimes even without regard to its possible physical meaning, may reduce the duration of oscillations of the sequence h_i when compared with the trace section s_i. Such a favourable case occurs when u and s_i resemble each other rather closely. In an ideal case of perfect similarity between u and s_i except for an amplitude factor and time delay, the corresponding h_i becomes an impulse, the height of which determines the amplitude of the phase relative to amplitudes of other phases. Location of the impulse in time defines the time delay of the ith phase. When $u = s_i$, for all i, then the original trace, s, is decomposed into a sequence of unit impulses indicating arrival times of individual phases. Negative unit impulses indicate reversals of seismic phases. On the other hand, when attenuation, scattering, etc. significantly modify the form of individual phases, then a suitable trace decomposition may be obtained only by means of a time-varying source wavelet. Accordingly, the process itself is then a *time-varying deconvolution* (Clarke, 1968).

Researchers have large freedom in selecting the most convenient form of the sequence u. Optimum forms vary from case to case. Robinson (1957) in his statistical approach uses *Ricker wavelets* (Ricker, 1940). Rice (1962) reduced the duration of original wavelets by more than 50% also by employing the Ricker wavelet. Howell et al. (1967) assumed high similarity between shapes of the first and later arrivals. Their approach helps, even if not unambiguously, to identify hidden pP-onsets. White and Mereu (1972) applied the concept of trace decomposition in deconvolving the refraction seismograms from underwater explosions. They derived the form of the source wavelet from the physical parameters of the source. Klíma and Kulhánek (1970) made use of an analytical *Berlage pulse*:

$$u(t) = t^{\alpha} e^{-\beta t} \sin \omega_0 t \qquad \text{for } t \geq 0$$

the form of which can be basically changed by the choice of parameters α, β, ω_0. Besides pP- and PcP-onsets from a teleseismic event several unidentified arrivals were revealed from the decomposed trace. This is rather typical for the decomposition processing. Due to the high noise sensitivity of the method, false onsets, corresponding to noise, unidentified phases, digitization errors, etc. may become very distinct on the transformed record.

Peacock and Treitel (1969) discussed the process of deconvolution in terms of optimum prediction filters with variable prediction times. Since prediction times control the length of the output from the deconvolution filter it is possible, within certain limits, to produce output traces with desired resolution. A qualitatively different approach to the deconvolution problem, the use of Kalman filters mentioned in Section 4.3, has been described by Bayless and Brigham (1970), Ott and Meder (1972) and Crump (1974).

Chapter 6

MULTIDIMENSIONAL FILTERS

The use of groups of identical seismometers, so-called *seismic arrays*, is not new in seismology. Profiles of seismometers have been used for seismic-reflection measurements since the 1920's. However, application of array techniques to earthquake seismology is new and did not start before the early 1960's. First, traces from individual sensors were combined to improve the signal/noise ratio in order to discriminate between earthquakes and low-yield underground nuclear explosions, as proposed by the Geneva Conference of Experts in 1958. Today, large arrays of apertures 100 km and more are also employed in event location, studies of the earth's structure, noise properties, etc.

Generally speaking, the *array design*, i.e. the geometrical pattern of a certain number of identical sensors in a given area, is based upon the idea that the signals passing across the array are coherent over the array area whereas the noise, in general, is not. The coherent and incoherent propagation is utilized in the improvement of the signal/noise ratio (Section 6.4). As will be mentioned later in this chapter, some types of seismic noise may also propagate coherently across the array. Various coherent signals and noise travel across the array site with different horizontal velocities and therefore may be separated from each other by employing so-called *velocity filtering* (Section 6.5).

Combining individual sensor outputs in the proper way improves the signal/noise ratio and enhances signals with a certain horizontal propagation velocity and arrival direction. Several essential methods, namely *straight summation* (SS), *delay-and-sum* (DS), *weighted delay-and-sum* (WDS) and more sophisticated *filter-and-sum* (FS) *techniques* are described in more detail in the following sections. The first three methods, SS, DS and WDS, are sometimes collectively called *beamforming*. The FS method may be viewed as weighting in the frequency domain, where different weights are applied to narrow frequency bands of the processed signal. Broadly speaking, an array system is a multiple-input system. Arrays discussed are *horizontal arrays* with linear processing techniques applied to outputs of individual sensors (the UK method, mentioned in Section 6.4, is an exception).

Some of the techniques may be directly applied as wavenumber–wavenumber filtering to process e.g. gravity or magnetic data.

6.1 SS TECHNIQUE

The straight summation of outputs of individual array sensors (see Fig. 6.1a) is the simplest array processing technique. Consider an array with M identical seismometers distributed arbitrarily on the horizontal plane of the earth's surface. No additional filtering systems are used in any of the M recording channels. Seismometer positions, with respect to the array origin, are specified by vectors, r_m, or by their polar coordinates (r_m, α_m) where $m = 1, 2, ..., M$ (see Fig. 6.2a). Coherent signal components travel across the array site with an apparent (move-out) velocity V, so that except for vertically propagating signals, phase shifts φ_m will occur at individual sensor outputs. When not mentioned otherwise, we shall assume the crustal structure to be identical beneath all sensors. Consider a stationary sinusoidal signal approaching the array site. Taking the array center as the reference point, we

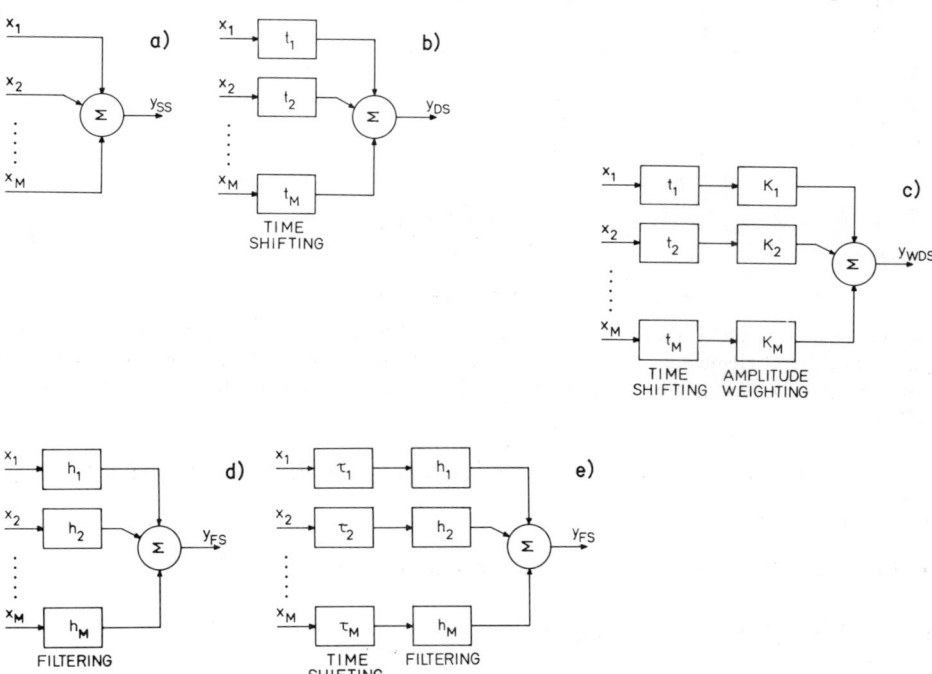

Fig. 6.1. Block-diagram representations for basic array (M sensors) processing techniques: (a) straight summation; (b) delay-and-sum; (c) weighted delay-and-sum; (d) high-pass velocity filtering; (e) band-pass velocity filtering.

have the reference signal $\sin(2\pi f t)$. As seen in Fig. 6.2a signals with the horizontal arrival direction θ and *horizontal wavelength* $\lambda = 1/k = V/f$ have phase shift at the mth sensor related to the array origin given by:

$$\varphi_m = (2\pi/\lambda) r_m \cos(\theta - \alpha_m) = 2\pi r_m k \cos(\theta - \alpha_m) \quad \text{for } m = 1, 2, ..., M \quad [6.1]$$

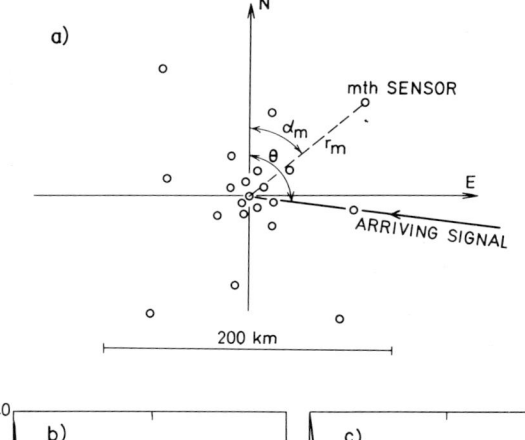

Fig. 6.2. Experimental LASA and cross-section, $\theta = N97°E$, of amplitude responses: (a) configuration of the 21-element array and the array coordinate system; (b) amplitude response for the 21-element array; (c) amplitude response for the 525-element compound array. After Lacoss (1965).

where k is the signal *horizontal wavenumber*. The mth output may be expressed as:

$$y_m = \sin(2\pi ft + \varphi_m)$$

and the straight summation of all M array outputs is:

$$y_{SS} = \sum_{m=1}^{M} \sin(2\pi ft + \varphi_m) = A \sin(2\pi ft + \varphi)$$

For the amplitude, A, we have relations:

$$A \sin \varphi = \sum_{m=1}^{M} \sin \varphi_m$$

$$A \cos \varphi = \sum_{m=1}^{M} \cos \varphi_m$$

$$A^2 = \left(\sum_{m=1}^{M} \cos \varphi_m \right)^2 + \left(\sum_{m=1}^{M} \sin \varphi_m \right)^2$$

The normalized amplitude response is the amplitude A divided by the number of sensors and can be expressed as a function of k and θ:

$$|y(k, \theta)| = \frac{|y_{ss}|}{M} = \frac{1}{M}\left[\left(\sum_{m=1}^{M} \cos \varphi_m\right)^2 + \left(\sum_{m=1}^{M} \sin \varphi_m\right)^2\right]^{1/2} \qquad [6.2]$$

Similarly, for the normalized power response we have:

$$|y(k, \theta)|^2 = \frac{|y_{ss}|^2}{M} = \frac{1}{M}\left[\left(\sum_{m=1}^{M} \cos \varphi_m\right)^2 + \left(\sum_{m=1}^{M} \sin \varphi_m\right)^2\right] \qquad [6.3]$$

For a given geometrical array configuration, responses [6.2] and [6.3] are two-dimensional functions of the wavenumber k and the angle θ. For the analysis of array performance it is useful to represent $|y(k, \theta)|$ or $y|(k,\theta)|^2$ by contour plots in the (k, θ) plane. To display the independent variables, polar diagrams with radius equal to k and azimuth equal to θ are used. Birtill and Whiteway (1965) present the power responses of straight summations for circular, symmetric cross, L-shaped and triangular arrays. Another example may be found in Burg (1964). It is also possible to diagram the cross-sections of array responses. This converts the two-dimensional function, say, $|y(k, \theta)|$ into one-dimensional $|y(k)|$ with θ as a parameter.

Broadly speaking, we require an array pattern which provides maximum response at and close to the origin of the (k, θ) plane, i.e. for $k \to 0$ and uniformly low response within a certain range of k around the origin. The range of low response is specified by the properties of noise and signals to be suppressed. When the array response satisfies the requirements mentioned, seismic signals with sufficiently small wavenumbers may be effectively separated from noise with higher wavenumbers via the simple straight summation. It follows from [6.1] that for vertically propagating signals, i.e. $k = 0$ and $\lambda = \infty$, outputs of all M sensors are in phase, $\phi_m = 0$ for $m = 1, 2, \ldots, M$, and the straight summation gives the maximum output irrespective of θ. As the wavelength decreases (k increases) from its infinite value, the summed output decreases rather rapidly. However, after passing a certain wavelength value the response shows a number of side lobes for further decreasing wavelengths. For any finite λ the response is also θ-dependent. Details obviously depend upon the geometrical configuration and the number of array sensors. An example is shown in Fig. 6.2a where a 21-element experimental Large Aperture Seismic Array (LASA) with an aperture of about 200 km is shown. The gain characteristics displayed in Fig. 6.2b have been evaluated along an azimuth $\theta = $ N97°E.

The straight summation method may also be applied to one-channel systems recording repeated signals. This concept is utilized e.g. in shallow seismic exploration where signals from repeated impacts are recorded by one geophone. Summation of aligned traces enhances highly repetitive signal components relative to unrepetitive noise components. The total enhancement increases with an increasing number of repeating traces. To obtain a better signal reproduction, individual traces may be prefiltered using optimum Wiener filters, for example, prior to the summation (Meyerhoff, 1966).

As a special application of the SS technique, let us consider an array of M sensors equally spaced along a straight line to form a *linear* or *one-dimensional array*, pointed towards the shot point or earthquake epicentre. It will be assumed that plane wavefronts approach the array site with a velocity v and successively excite individual array sensors. On adjacent sensors, the arrival times are τ sec apart. The quantity τ is called the *signal move-out* or *step-out*.

As follows from Fig. 6.3, the excitation propagates with an *apparent velocity* $V = v/\sin \psi$ on the earth's surface along the array line. The velocity V depends upon the arrival direction and the local structure beneath the array. It approaches the value v for horizontally propagating wavefronts and is infinitely large for vertically propagating wavefronts. Accordingly, we may define the *apparent wavelength*, $\lambda = V/f$, and the horizontal wavenumber, $k_a = 1/\lambda$, along the array.

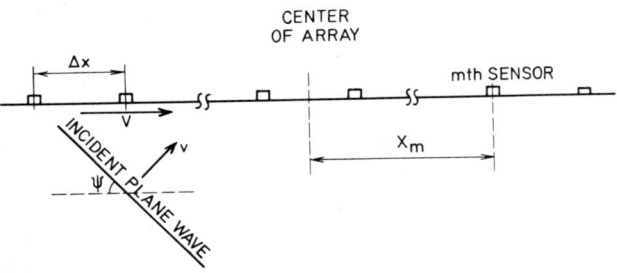

Fig. 6.3. Sensor spacing and apparent velocity, V, of a wavefront propagating along a linear array.

Let us assume a simple stationary waveform, $A \sin 2\pi ft$, approaching the sensor nearest to the source. The corresponding plane wavefronts travel across the array site at an angle ψ with the horizontal. Individual sensors will record successively delayed waveforms $A \sin 2\pi ft$, $A \sin 2\pi f(t + \tau)$, ..., $A \sin 2\pi f [t + (M - 1)\tau]$ (see also Fail and Layotte, 1970). The last waveform mentioned belongs to the most distant sensor. The step-out, τ, is constant across the array and may be expressed (see Fig. 6.3) as:

$$\tau = (\Delta x \sin \psi)/v = \Delta x/V$$

Making use of the Euler formula and of the formula for the sum of geometrical series, it may be shown that the amplitude of the summed output of the array, divided by the total number of sensors, is KA, where:

$$K = \frac{\sin M2\pi f\tau/2}{M \sin 2\pi f\tau/2}$$

The amplitude response, being dependent upon the number of sensors, frequency of arriving signals and signal move-out, may be utilized in obtaining various array characteristics. It is obvious that the signal move-out may be expressed in terms of the sensor spacing and apparent velocity which, in turn, may be defined via the apparent wavelength or wavenumber. It is also possible to include the angle ψ as an independent variable. Hales and Edwards (1955) present a series of diagrams showing the dependence of the factor K upon the angle ψ, so-called *directional selectivity* in the vertical plane of the linear array. They also studied the frequency selectivity for signals arriving at a given angle ψ and for given spacing Δx. Response curves as functions of the total array length, signal wavelength and number of sensors have been worked out by Lombardi (1955). Curves for 2, 4, 9 and an infinite number of sensors show the decrease of side-lobe amplitude with an increasing number of seismometers. Smith (1956) passed the summed output of a 10-element-array through a band-pass frequency filter and studied the combined array response as a function of frequency and wavenumber. Verma and Roy (1970) described a graphical approach in constructing the array response. Vinnik (1963) derived amplitude-response relations for arrays with sensors horizontally located on a regular rectangular grid.

6.2 BASIC REQUIREMENTS FOR ARRAY PATTERNS

The design of the array configuration and the choice of the number of array elements is a rather complicated problem. A satisfactory solution is usually obtained by some kind of the trial-and-error method using the long-term statistics of recorded phenomena (Lacoss, 1965). Behaviour of several uniform configurations has been discussed by Haubrich (1968). Since detailed discussion exceeds the scope of the present book, we summarize only several basic principles. For more details the reader is referred e.g. to Green (1965), Lacoss (1965) or to other sources.

The aperture of the array defines the *wavenumber resolution*, that is, the width of the peaks in the response diagrams. High resolution is necessary to be able to distinguish between signals with close (k, θ) parameters. For example, peak widths of about 0.01 km^{-1} (compare with Fig. 6.2b) require a minimum aperture of about 100 km. The choice of the number of sensors is ruled by several opposing requirements. An increase of M decreases the average level of undesired side lobes while it obviously increases the array cost and processing complexity. The side-lobe structure is to some extent controlled by weighting factors attached to individual sensors (Green et al., 1965). It seems that the *spatial aliasing* phenomenon is decisive for the number of seismometers to be used. *Spatial sampling* in a regular space grid will generate a *wavenumber aliasing*, i.e. the array response becomes a periodic function of wavenumber with a period $1/\Delta x$, where Δx is the sensor spacing. The highest resolvable wavenumber is $k_N = 1/2\Delta x$. The spatial aliasing is analogous to that of frequency aliasing discussed in Chapter 1. For example, when we choose $\Delta x = 10$ km then $k_N = 0.05$ km^{-1} which is inadmissibly low. Note that P-waves travelling across the array with an apparent velocity of about 8 km/sec and higher and with dominant frequencies close to 1 cps will have a wavenumber $k \leq 0.13$ km^{-1}. This means that the sensor spacing $\Delta x = 10$ km aliases a large part of the P-wave energy. To minimize the effect of the spatial aliasing, sensor spacing of about 1 km or less is required. But for the aperture mentioned above this would require about 10^4 sensors. Besides, dense sensor spacing naturally increases undesired noise coherence between channels. A less costly and more elegant way to avoid aliasing problems is the usage of sufficiently irregular array configurations. For LASA, a satisfactory solution has been achieved by coupling together 21 *subarrays*, each 7 km in diameter and having 25 sensors, into a so-called *compound array* having altogether 525 sensors and about 200 km in diameter. The configuration of the compound array corresponds to that displayed in Fig. 6.2a where each of the 21 sensors represents a 25-element subarray. The minimum element spacing within the subarray was chosen as 0.25 km. The influence of the number of sensors upon the level of side lobes is clearly seen in Figs. 6.2b and 6.2c. On the other hand, note that whereas the difference in installation of a 21-element array vs. a 525-element array is tremendously large, there is no significant response improvement for, say, $k \leq 0.10$ km^{-1} (compare Figs. 6.2b and 6.2c).

Besides its configuration and number of sensors, the decisive factor for the proper functioning of a seismic array is the degree of similarity between signals recorded by individual array seismometers. Here, the geological structure beneath the array is of greatest importance. Homogeneous structure with horizontal layering is required.

6.3 DS AND WDS TECHNIQUES

The straight-summation method, outlined in Section 6.1, provides most effective enhancement for signals with infinite or very long wavelengths. However, all signals do not satisfy this condition. It is not difficult to modify the array response so as to move the response peak to any desired position in the (k, θ) plane. Assuming that the signal to be enhanced has parameters (k_1, θ_1) and that the corresponding apparent velocity is V_1, let us apply to individual sensors a phase shift:

$$\gamma_m = 2\pi r_m k_1 \cos(\theta_1 - \alpha_m) \quad \text{for } m = 1, 2, \ldots, M$$

Thus, an arbitrary signal (k, θ) recorded by the mth sensor has a phase shift related to the array origin which is given by:

$$\rho_m = \phi_m - \gamma_m = 2\pi r_m [k \cos(\theta - \alpha_m) - k_1 \cos(\theta_1 - \alpha_m)] \quad \text{for } m=1, 2, \ldots, M \quad [6.4]$$

It follows immediately from [6.4] that for signals $(k = k_1, \theta = \theta_1)$, all M sensor outputs are in phase with $\rho_m = 0$ for any m, and the array response is a maximum.

Equation [6.4] expresses the difference between the SS and DS techniques.

While in the former case we carry out a straight summation of individual outputs, in the latter case we perform a *delayed summation*. Delay-and-sum method implies a time shifting of the mth output by:

$$t_m = [r_m \cos(\theta_1 - \alpha_m)]/V_1 \quad \text{for } m = 1, 2, \ldots, M \tag{6.5}$$

prior to the summation. The time shift t_m corresponds to the phase shift γ_m. The block-diagram representation for the DS technique is shown in Fig. 6.1b.

Introduction of time delays to individual outputs is called *array steering* (*tuning*), sometimes also *array alignment*. Aligned arrivals correspond to a fictitious wavefront propagating vertically beneath the array, while the DS technique steers the array in the (k, θ) plane, moving the response maximum from $(k = 0, \theta = 0)$ to a new position $(k = k_1, \theta = \theta_1)$. The apparent velocity and azimuth selectivity of the array response is then evident. A somewhat more sophisticated alteration of the DS method including also two nonlinear operations has been outlined by Kanasewich et al. (1973).

There are several methods available for choosing array alignment. The simplest one is the visual alignment. Signals are shifted in time according to the measured differences between the corresponding arrival times. In fact, any highly correlated peak or trough on the nondispersive wave group and not only the first onset may be used. Kulhánek (1973) showed that for good quality short-period records, one may expect the error of visual steering to fall within the range ± 0.1 sec. Special care has to be taken for signals with more or less periodic form. If this is the case, then alignment due to noncorresponding peaks or troughs, so-called *velocity aliasing*, may occur.

For a more sophisticated alignment the time delay between two adjacent sensors may be estimated a priori from [6.5], assuming plane wavefronts and constant apparent velocities across the array. Dean (1965) pointed out that due to the possible velocity anomalies beneath individual seismometers, this technique may not be successful.

Finally, it is possible to cross-correlate all traces against the reference trace and read the time delays corresponding to maxima in individual cross-correlations (see e.g. Dean, 1965 or Kulhánek, 1973). The method itself can be applied only to traces with high signal/noise ratio because otherwise the noise is aligned rather than the signals. On the other hand, the cross-correlation takes into account the possible velocity anomalies mentioned above.

Provided that the signal at the reference point has a form $\sin 2\pi f t$ and utilizing phase shifts defined in [6.4], the output of the mth sensor becomes:

$$y_m = \sin(2\pi f t + \rho_m)$$

and the amplitude response may be written as:

$$\frac{|y_{DS}|}{M} = \frac{1}{M} \left| \sum_{m=1}^{M} y_m \right|$$

$$= \frac{1}{M} \left\{ \left[\sum_{m=1}^{M} \cos(\phi_m - \gamma_m) \right]^2 + \left[\sum_{m=1}^{M} \sin(\phi_m - \gamma_m) \right]^2 \right\}^{1/2} \qquad [6.6]$$

It is not difficult to complete the list of array processing techniques by adding the weighted delay-and-sum technique indicated in Fig. 6.1c. Including weighting factors, K_m for $m = 1, 2, \ldots, M$, to each of the array channels, so-called *tapered arrays*, provides the mth output:

$$y_m = K_m \sin(2\pi ft + \rho_m)$$

For the amplitude response we have:

$$\frac{|y_{\text{WDS}}|}{M} = \frac{1}{M} \left\{ \left[\sum_{m=1}^{M} K_m \cos(\phi_m - \gamma_m) \right]^2 + \left[\sum_{m=1}^{M} K_m \sin(\phi_m - \gamma_m) \right]^2 \right\}^{1/2}$$

As in the preceding case, $|y_{\text{WDS}}|$ is a function of wavenumber and horizontal arrival direction. Properly chosen K_m coefficients improve the output signal/noise ratio and suppress the undesired side lobes of the array response. Savit et al. (1958) present results from a tapered array of 21 sensors. After introducing weighting coefficients, the array shows a significant improvement in side-lobe amplitudes when compared with an array with equal weighting for all its sensors (for single-frequency input signals the amplitude weighting and phase shifts are the only possibilities to minimize the side lobes). The improvement of the signal/noise ratio gained by applications of DS and WDS techniques is treated in more detail in the following section.

6.4 IMPROVEMENT OF THE SIGNAL/NOISE RATIO DUE TO DS AND WDS TECHNIQUES

Array alignment brings signal components of individual sensors into time coincidence. This does not affect the noise components provided that these are uncorrelated among seismometers. Let us investigate the improvement of the signal/noise ratio which may be achieved by applying the DS and WDS processing techniques.

Consider an array of M identical seismometers spaced over a distance comparable with wavelengths of signals under interest. Consider further high signal coherence and negligible noise coherence across the array. We define the variance σ_x^2 of a digitized trace x_i as:

$$\sigma_x^2 = \frac{1}{N} \sum_{i=1}^{N} (x_i - \bar{x})^2$$

where N is the number of the points considered and $\bar{x}$ is the average trace amplitude. There are a number of different definitions in the literature of the signal/noise ratio, one example being shown in Section 4.2. Here, let the signal/noise power ratio of the mth trace be defined by:

IMPROVEMENT OF THE SIGNAL/NOISE RATIO

$$Q_m = \frac{\sigma_{sm}^2}{\sigma_{nm}^2} \quad \text{for } m = 1, 2, \ldots, M \qquad [6.7]$$

where σ_{sm}^2 and σ_{nm}^2 are the signal and noise variances respectively.

Traces recorded by individual sensors may be viewed as sums of the coherent signal and random noise components. Hereafter, it will be assumed that signals and noise have zero means. For M seismometers we have:

$$x_{1i} = s_{1i} + n_{1i}$$
$$x_{2i} = s_{2i} + n_{2i}$$
.
.
.
$$x_{Mi} = s_{Mi} + n_{Mi} \qquad \text{for } i = 1, 2, \ldots, N$$

The variance of the sum of aligned signal components, i.e. the variance of the signal at the array output, becomes:

$$\sigma_s^2 = \frac{1}{N} \sum_{i=1}^{N} (s_{1i} + s_{2i} + \ldots + s_{Mi})^2$$

Assuming perfect signal similarity, i.e. $s_{1i} = s_{2i} = \cdots = s_{Mi}$ for any i, we write:

$$\sigma_s^2 = \frac{M^2}{N} \sum_{i=1}^{N} s_i^2 \qquad [6.8]$$

where s_i is the digital signal recorded by any of the M sensors. For the sum of random output noise we obtain:

$$\sigma_n^2 = \frac{1}{N} \sum_{i=1}^{N} (n_{1i}^2 + n_{2i}^2 + \ldots + n_{Mi}^2)$$

Considering the simplest case of equal noise levels across the entire array, i.e.:

$$\sum_{i=1}^{N} n_{1i}^2 = \sum_{i=1}^{N} n_{2i}^2 = \ldots = \sum_{i=1}^{N} n_{Mi}^2$$

the variance of the array output noise component yields:

$$\sigma_n^2 = \frac{M}{N} \sum_{i=1}^{N} n_i^2 \qquad [6.9]$$

where n_i again is the digitized noise component from any of the M sensors. Introducing the new variances [6.8] and [6.9] into [6.7] we determine the signal/noise power ratio of a summed output as:

$$Q_{DS} = \frac{\dfrac{M^2}{N} \sum_{i=1}^{N} s_i^2}{\dfrac{M}{N} \sum_{i=1}^{N} n_i^2} = MQ \qquad [6.10]$$

Under these conditions, the delay-and-sum technique provides an improvement in Q which is equal to the number of sensors when compared with a single sensor. Note that [6.10] specifies the improvement of a power ratio. When the improvement of an amplitude ratio is of interest, a value $\sqrt{M}$ rather than M should be used (see e.g. Koopmans, 1961).

When the noise is correlated or partly correlated among sensors, expression [6.10] must be modified accordingly. Denham (1963) assumes the same noise level at all sensors and calculates the improvement of the signal/noise amplitude ratio as:

$$\frac{M}{[M + (M^2 - M)\bar{\rho}]^{1/2}}$$

where $\bar{\rho}$ is the average of noise correlation coefficients. For example, in a 3-element array $\bar{\rho} = (\rho_{12} + \rho_{13} + \rho_{23})/3$. Denham's signal/noise ratio differs slightly from that given by [6.7], nevertheless, it is obvious that for completely random noise, $\bar{\rho} = 0$, the improvement (in amplitudes) is again equal to $\sqrt{M}$.

Constant signal and noise levels over the entire array area as considered above is a rather ideal case. Differences in noise levels, coupling and instrumental effects, fine geological structure beneath array sensors, etc., all contribute to variations in the signal/noise ratios among individual traces. If this is the case, a simple DS technique does not provide the best solution. In addition to steering delays, different weighting factors must be applied to each sensor prior to summation to obtain the best Q improvement. Block-diagram representation of this weighted delay-and-sum technique is shown in Fig. 6.1c.

Let us apply amplitude coefficients $K_1, K_2, ..., K_M$ to each$_M$ of the M seismometers. The corresponding outputs are then:

$$x_{1i} = K_1(s_{1i} + n_{1i})$$

.
.
.

$$x_{2i} = K_2(s_{2i} + n_{2i})$$
$$x_{Mi} = K_M(s_{Mi} + n_{Mi}) \qquad \text{for } i = 1, 2, ..., N$$

Again assuming perfect signal similarity across the array, i.e. $s_{1i} = s_{2i} = ... = s_{mi}$, the variance of the weighted and summed signal components becomes:

$$\sigma_s^2 = \frac{1}{N} \sum_{i=1}^{N} (K_1 s_{1i} + K_2 s_{2i} + ... + K_M s_{Mi})^2 = \left(\sum_{m=1}^{M} \sigma_{sm} K_m \right)^2$$

where σ_{sm}^2 is the signal variance of any of the M traces. The variance of the array output uncorrelated noise is:

$$\sigma_n^2 = \frac{1}{N} \sum_{i=1}^{N} (K_1 n_{1i} + K_2 n_{2i} + ... + K_M n_{Mi})^2 = \sum_{m=1}^{M} \sigma_{nm}^2 K_m^2$$

Consequently, the signal/noise power ratio of the array output is expressed as:

$$Q_{\text{WDS}} = \frac{\sigma_s^2}{\sigma_n^2} = \frac{\left(\sum_{m=1}^{M} \sigma_{sm} K_m\right)^2}{\sum_{m=1}^{M} \sigma_{nm}^2 K_m^2} \qquad [6.11]$$

In order to find the amplitude coefficients K_m that minimize Q_{WDS}, we equate $\delta Q_{\text{WDS}}/\delta K_m$ to zero and check the sign of the second derivative. From this it follows that the optimum coefficients satisfy the condition:

$$K_m = K\sigma_{sm}/\sigma_{nm}^2$$

where K is an arbitrary constant (see also Birtill and Whiteway, 1965). Introducing the optimum values K_m into [6.11] we have:

$$Q_{\text{WDS}} = \sum_{m=1}^{M} \frac{\sigma_{sm}^2}{\sigma_{nm}^2} \qquad [6.12]$$

That is, when optimum amplitude weighting coefficients are employed, the WDS technique provides a Q_{WDS} ratio which is equal to the sum of individual signal/noise power ratios, provided that the condition of perfect signal similarity across the array holds.

Consider, as a simple example, a three-element array with the following signal and noise variances:

$\sigma_{s1} = \sigma_1 \qquad \sigma_{s2} = \sigma_1 \qquad \sigma_{s3} = \sigma_1$

$\sigma_{n1} = \sigma_2 \qquad \sigma_{n2} = 2\sigma_2 \qquad \sigma_{n3} = 3\sigma_2$

The three optimum weighting coefficients are:

$K_1 = \sigma_1/\sigma_2^2 \qquad K_2 = \sigma_1/(4\sigma_2^2) \qquad K_3 = \sigma_1/(9\sigma_2^2)$

For this weighting, [6.12] yields $Q_{\text{WDS}} = 1.36\ \sigma_1^2/\sigma_2^2$. With respect to the first seismometer (the best signal/noise power ratio) the WDS technique provides an improvement by a factor of 1.36. If the sensor outputs are not individually weighted, i.e. $K_1 = K_2 = K_3 = 1$, the resultant ratio $Q_{\text{WDS}} = 0.64\ \sigma_1^2/\sigma_2^2$. Thus, the direct application of the DS technique would in fact lead to a decrease of the signal/noise power ratio when compared with the single output of the best seismometer.

Naturally, in choosing the coefficients K_m it is possible to consider requirements other than the maximum signal/noise ratio. For example, Holzman (1963) and Petrov (1963) employ Chebyshev polynomials to construct the best (in least-squares sense) response approximation of a linear tapered array with M sensors. For a specified main lobe width, the *optimized Chebyshev array* yields minimum amplitude and equiripple response in the rejection band. The limiting case of zero side lobes in the Chebyshev array results in the *binominal array* with coefficients

$K_{i+1} = \binom{n}{i}$, where $n = M - 1$ and $i = 0, 1, \ldots, M - 1$. The price to be paid for the zero response in the rejection band is the rather broad main lobe. Further improvement of the array performance, derived from a Chebyshev array, is described by Schoenberger (1970).

We remark here that improvements defined in [6.10] and [6.12] provide theoretical values which may differ significantly from those observed. Note that an increase of M in a given area will most likely increase the noise coherence among sensors, thus decreasing Q_{DS} and Q_{WDS}. Hartenberger and Van Nostrand (1972) show that with increasing number of sensors in a proportionately increasing area (i.e. constant sensor spacing), the improvement of the signal/noise ratio increases asymptotically to a practical maximum. This means that, after reaching a certain level of Q_{DS} and Q_{WDS} any further improvement will be very costly. The physical explanation for this effect probably is the decreasing signal coherence within the increasing array area.

In some applications, especially when the array sensors form two separate groups of seismometers (such as cross arrays, L-shaped arrays, etc.), a cross-correlation technique may be successfully employed (see e.g. Ryall, 1964; Somers and Manchee, 1966; Iyer, 1968; King et al., 1973). Cross-correlating the outputs of both groups against each other gives an improved signal/noise ratio (Birtill and Whiteway, 1965). The method is sometimes called the UK method.

6.5 VELOCITY FILTERING

This section deals mostly with high- and band-pass velocity filtering. By making use of the more sophisticated filter-and-sum technique, multi-channel velocity filters are constructed for one- as well as for two-dimensional arrays. Velocity filters provide a means to discriminate signals from noise or other undesired signals due to their different apparent velocities. Thus, the use of velocity filters makes it possible to enhance even signals occupying the same frequency ranges as noise or undesired signals do. As will be shown in the following subsections, velocity filtering is performed by multi-dimensional filters included in the array processing scheme. Characteristics of these filters are expressed by two- and three-dimensional functions covering the whole array site.

The application of multi-dimensional velocity filters should not be confused with a possible use of one-dimensional frequency filters designed separately for individual array channels.

6.5.1 *Signal and noise characteristics*

In the following list we briefly summarize the essential characteristics of desired and undesired seismic waveforms.

Teleseismic signals. Apparent velocity of arriving signals depends upon the wave type, source location and structure beneath the recording site. Roughly speaking, observed apparent velocities will lie above 8 km/sec for P-waves and above 4 km/sec for S-waves. Whereas the former wave type shows predominant frequencies around 1 cps (recorded by SP instruments) the latter has usually lower peak frequency. Velocities of surface waves, which are frequency-dependent, may reach values up to about 4.5 km/sec. It is assumed that teleseismic signals propagate coherently across the array.

Microseismic noise. This type of noise, also called ocean microseisms, propagates primarily as Rayleigh waves with velocities from about 2.5 km/sec to 4 km/sec. Dominant frequencies occupy a broad low-frequency range. Depending upon the distance to the source and sensor spacing, some portion of microseisms may propagate coherently across the array.

Incoherent noise. Various human activities (factories, traffic, construction), action of wind, smaller water basins, etc., generate high-frequency noise above 1 cps. Corresponding apparent velocities vary for different sources, but in general have low values, around 1 km/sec. For sensor spacing larger than several kilometers it is most likely that the high-frequency noise propagates incoherently across the array, irrespective of whether the location of the noise source is inside or outside the array pattern.

Signal-generated noise. This is noise generated by the desired signal itself in the vicinity of the sensor as multiple reflections and mode conversions. It occupies a wide range of apparent velocities, depending upon the local structure and it may propagate coherently across the array.

A variety of sources generate a variety of seismic signals and noise types. To state commonly valid characteristics is rather difficult and for every particular case the above list may be completed accordingly.

6.5.2 *Velocity filtering by means of one-dimensional arrays*

Fig. 6.4 shows an idealized distribution of teleseismic P-wave signals and the coherent noise, mentioned in the preceding subsection, in the frequency vs. wavenumber plane. Observed P-waves with apparent velocities in the range from 8 km/sec to infinity occupy the triangular zone labeled "signal zone". All kinds of coherent noise with apparent velocities between 2.5 km/sec and 4 km/sec are distributed within the wedge marked "coherent noise".

Consider a one-dimensional array as depicted in Fig. 6.3 and horizontal wave propagation parallel to the line of sensors. An adequate separation of the signal from the ambient noise may be achieved by adopting an operator based upon velocity discrimination. Using velocity discrimination or velocity filtering, we define the pass and rejection zones by straight lines (no dispersion considered) through the origin in the (f, k) plane. Besides sampling, there are no other limitations upon frequency and wavenumber intervals. Spatial sampling by equally spaced sensors results in a periodic amplitude response in wavenumber. In Fig. 6.4

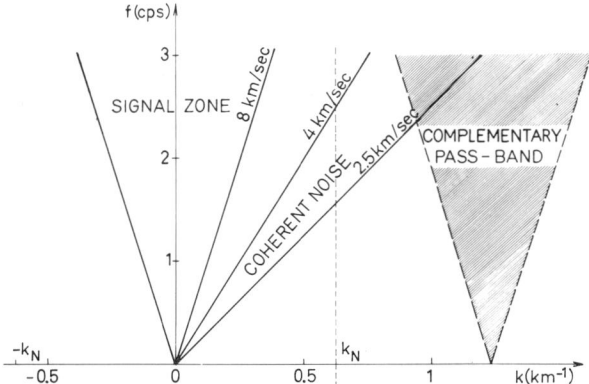

Fig. 6.4. Idealized distribution of the signal and noise components in the frequency vs. wavenumber plane. Space sampling $\Delta x = 0\,8$ km/sec and time sampling $\Delta t = 0\,1$ sec have been used.

the shaded wedge represents the first "positive" complementary (aliased) pass band. It follows from the figure that due to the first and higher aliased bands, the operator may pass through some part of energy even from the rejected velocity band. The amount of this energy can be controlled, to some extent, by k_N, i.e. by the sensor spacing as mentioned in Section 6.2. Provided that the coherent noise occupies a region in the (f, k) plane that differs from that of the signal, the velocity filtering performs a perfect discrimination against the coherent noise.

Both the transfer and impulse-response functions of a velocity filter are two-dimensional functions. The former is a function of frequency and wavenumber whereas the latter depends upon time and location within the array. Thus, velocity filtering by means of linear arrays, represents a two-dimensional multi-channel (M sensors) processing technique. It differs from the conventional prefilter-and-sum scheme in that the trace from each sensor is filtered by a different appropriately constructed filter. The prefilter-and-sum processing uses a set of individually designed one-dimensional frequency filters. Velocity filtering as a more complex technique employs two-dimensional, i.e. frequency–space interconnected filters. The mathematical background of the two-dimensional velocity filtering has been adequately treated in several papers, see e.g. Fail and Grau (1963), Embree et al. (1963), Wiggins (1966) or Nakhamkin (1969). The possibility of employing Green's theorem in constructing two-dimensional filters specified in the f–k domain has been discussed by Ford (1967). Our development generally follows that of Embree et al. (1963).

Let us assume that it is desired to pass waveforms with wavenumbers within the range $-|f|/V < k < |f|/V$. Outside this wavenumber range all waveforms are rejected. The two-dimensional transfer function is then defined as:

$$H(f, k) = \begin{cases} 1 & -\dfrac{|f|}{V} \leq k \leq \dfrac{|f|}{V} \\ 0 & \text{elsewhere} \end{cases} \qquad [6.13]$$

The general shape of $H(f,k)$ with pass and rejection zones (primary pass band only) is diagrammed in Fig. 6.5 (left part). Due to the shape of the transfer function, the

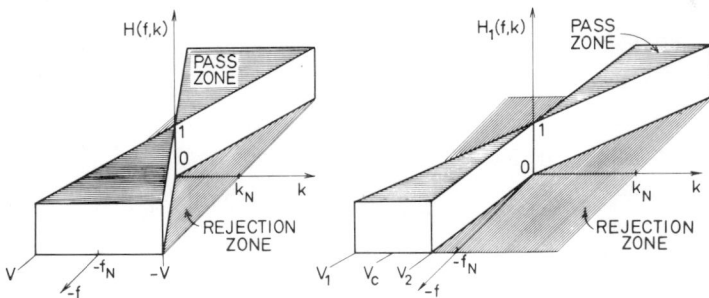

Fig. 6.5. Ideal high-pass (left part) and band-pass (right part) velocity-transfer function. Only primary pass zones depicted.

filtering itself is also called *pie-slice processing* (Embree et al., 1963). Some authors, see e.g. Treitel et al. (1967), prefer the name *fan filtering* or *fan-pass filtering* taken over from *"filtre en éventail"* of Fail and Grau (1963).

The time–space impulse response of the filter may be expressed in terms of the inverse two-dimensional Fourier transform of $H(f,k)$, see e.g. Båth (1974, pp. 63–66), as:

$$h(t, x) = \int_{-\infty}^{\infty} \int_{-\infty}^{\infty} H(f, k)\, e^{j2\pi(ft+kx)}\, df\, dk \qquad [6.14]$$

where t is time and x is position along the line of sensors. In other words, [6.14] provides the value of the unit-impulse response at any time instant, t, and for any sensor of the array.

Due to the periodicity of $H(f, k)$ in wavenumber we must limit ourselves to the resolvable wavenumber band only:

$$-k_N \leq k \leq k_N$$

It follows from the basic relation $f = Vk$, that for a given apparent velocity, V, and frequencies $|f| > Vk_N$ in the (f,k) plane, we leave the primary and enter the complementary pass zone. Therefore, the frequency limits for a meaningful transfer function are:

$$-f_N \leq f \leq f_N$$

where $f_N = Vk_N = V/2\Delta x$. Having established the value of the folding frequency,

f_N, the sampling period, Δt, which preserves all information within the frequency band $-f_N, f_N$ becomes:

$$\Delta t = 1/2f_N$$

The last equation also shows that the apparent cut-off velocity satisfies:

$$V = f_N/k_N = \Delta x/\Delta t$$

For any signal to pass through this velocity filter the apparent velocity must be equal to or higher than V, i.e. for a given sensor spacing the signal move-out, τ, must satisfy the condition:

$$\tau \leq \Delta t$$

In Fig. 6·4 values $\Delta x = 0.8$ km and $\Delta t = 0.1$ sec have been used.

Introducing the finite limits $\pm f_N, \pm k_N$ [6.14] is rewritten as:

$$h(t, x) = \int_{-f_N}^{f_N} \int_{-k_N}^{k_N} e^{j2\pi (ft + kx)} \, df \, dk \qquad [6.15]$$

Embree et al. (1963), Finetti et al. (1971) and Kanasewich (1973) among others present the solution of [6.15] for an even number of seismometers. Denoting by X_m the distance of the mth sensor from the center of the array and by $T_n = n\Delta t$ the nth sampling instant, they present the time–space operator [6.15] in a form:

$$h(T_n, X_m) = \{\pi^2 [(X_m/\Delta x)^2 - (T_n/\Delta t)^2]\}^{-1} \qquad [6.16]$$

As follows from [6.16], $h(T_n, X_m)$ is a real and even function in the (T_n, X_m) plane. Therefore, any velocity filter specified by [6.16] operates as a phase-distortionless system. The block-diagram representation of the high-pass velocity filter [6.16] is shown in Fig. 6.1d.

Equation [6.16] may be also visualized as a description of a set of interconnected frequency filters $H_m(f)$, where $m = 1, 2, \ldots, M$, included in each of the array channels. A system of equations defining the set of functions, $H_m(f)$, has been derived by Sengbush and Foster (1968). These filters, applied to respective array channels, perform a Wiener-sense optimum velocity filtering. Both pass as well as rejection filters are discussed. Galbraith and Wiggins (1968) extended the optimum multi-channel filtering to multiple-input–multiple-output systems.

Equation [6.13] defines a high-pass velocity filter which passes signals with apparent velocities of magnitude greater than V. Signals with lower velocities are rejected. Such a filter can easily be used for band-pass velocity filtering as well. Consider a velocity pass range from V_1 to V_2 with a center velocity V_c. The corresponding transfer function, $H_1(f, k)$, is depicted at the right side of Fig. 6.5. Let us rotate $H_1(f, k)$ counterclockwise about the origin. At a position when V_c becomes infinitely large, $H_1(f, k)$ is transformed into a high-pass-type transfer function similar to $H(f, k)$. Signals which propagate along the array with a velocity of V_c arrive at adjacent sensors at times τ_c apart. When rotating $H_1(f, k)$, $V_c \rightarrow \infty$ and

VELOCITY FILTERING

$\tau_c \to 0$. Thus, rotation in the f–k domain corresponds to time shifting in the time domain. After alignment of onsets, corresponding to the velocity V_c, function [6.16] may be employed as in Fig. 6.1e. In other words, time shifting and high-pass velocity filtering replace the band-pass velocity filtering. Examples of band-pass velocity filters may be found in Foster et al. (1964), Kanasewich (1973) and elsewhere.

To obtain the output sequence y resulting from velocity filtering, the M input sequences must be convolved with the two-dimensional operator h and summed:

$$y_n = \sum_{m=1}^{M} x_{mn} * h_{mn} \qquad [6.17]$$

where x_{mn} and h_{mn} are the input and impulse-response sequences respectively, m is the input channel index and n is an integer time index. The symbol * denotes time-domain convolution. The output corresponds to $X_m = 0$, i.e. to a filtered trace which would have been recorded at the center point of the array. Note that no steering delays are involved in the processing. It is assumed that the alignment is included in the filter performance.

Because of the spatial symmetry of $h(T_n, X_m)$, the two traces generated by a pair of sensors on both sides of the array and equally distant from its center are convolved by the same operator. It is convenient to introduce the sums $u_1 = x_1 + x_M$, $u_2 = x_2 + x_{M-1}, \ldots, u_{M/2} = x_{M/2} + x_{M/2+1}$ prior to the convolution indicated in [6.17]. Making use of the linearity property, the output sequence can be written as:

$$y_n = \sum_{m=1}^{M/2} u_{mn} * h_{mn}$$

and the total number of convolutions to be performed is reduced from M, in [6.17], to $M/2$. Further reduction of the number of convolution operations has been suggested by Treitel et al. (1967). They derived a recursion relation for the velocity filter of Embree et al. (1963) which requires only one convolution per output trace, irrespective of the total number of sensors involved. The results have also been extended to velocity-rejection filters.

The theoretical response function corresponding to the time-space impulse response, [6.16], may be determined via the two-dimensional Fourier transform (see e.g. Robinson, 1967b, p. 319) as:

$$H(f, k) = \sum_{m=-\infty}^{\infty} \sum_{\substack{m=-\infty \\ m \neq 0}}^{\infty} h(T_n, X_m) \exp[-j2\pi(fT_n + kX_m)]$$

In practical calculations only finite summations may be carried out and therefore $H(f, k)$ can be only approximated. It may be expected intuitively that the sharpness between the pass and rejection zones increases and the amplitude of side lobes decreases with an increasing number of array sensors (see also Cassano and Rocca, 1974). A formula relating the number of sensors per unit length, the cut-off wavenumber and the distance from the center of the array, has been presented by Krey and Toth (1973). In prospecting work, 6–12 sensors is usually a sufficient number. A true twelve-trace velocity-filter amplitude-response has been presented by Fail and Grau (1963) and by Embree et al. (1963). Other examples may be found in Treitel et al. (1967) or elsewhere. McClellan and Parks (1972) described the construction of equiripple fan filters based upon the Chebyshev approximation of a rectangular function.

6.5.3 Two-dimensional wavenumber filtering

Two-dimensional filtering is not limited to velocity filtering by means of seismic arrays. Two-dimensional spatial filters are frequently used e.g. in the interpretation of gravity and magnetic maps to differentiate between regional and local features. Basic principles of the two-dimensional wavenumber digital filtering are treated adequately e.g. by Clement (1973). Input data for these filters may be observations in the survey of an area conducted over a grid in the (x,y) plane of the earth's surface. The response functions of these filters are usually specified in terms of wavenumbers or other frequency parameters in the x- and y-directions (Byerly, 1965; Dampney, 1965; Black and Scollar, 1969; Clarke, 1969; Agarwal and Lal, 1970; Mufti, 1972; Lapina and Strakhov, 1973) rather than in the (f, k) plane.

When interpreting fine fluctuations appearing in gravity and magnetic maps *second-derivative* and *downward-continuation methods* are frequently used. Both methods may provide better material than the original maps. Consider the measurements $f(x, y)$ made on the earth's surface in a regular grid. Either of the two techniques mentioned may be visualized in terms of operators $h(x, y)$ or $H(k_x, k_y)$ applied to original data (Meskó, 1965, 1966; Darby and Davies, 1967; Ku et al., 1971; Bhattacharyya, 1972):

$$g(x, y) = h(x, y) * f(x, y) \quad \text{and} \quad G(k_x, k_y) = H(k_x, k_y) F(k_x, k_y)$$

where $g(x, y)$ represents a set of "improved" measurements, the symbol $*$ denotes convolution and capital letters denote Fourier transforms. In the wavenumber domain operators corresponding to the second-derivative and downward-continuation techniques have simple expressions (see e.g. Clarke, 1969) $H_1(k_x, k_y) = k_x^2 + k_y^2$ and $H_2(k_x, k_y) = \exp(z\sqrt{k_x^2 + k_y^2})$, respectively. Note, that $H_1(k_x, k_y)$ as well as $H_2(k_x, k_y)$ increase without limit as the independent variables increase. Consequently, the approach mentioned is applicable only when $F(k_x, k_y)$ converges faster than $H(k_x, k_y)$. In practice, some approximate formulae are usually employed instead of these exact expressions. It follows immediately from the above equations for $g(x, y)$ or $G(k_x, k_y)$ that both these techniques may be visualized as two-dimensional filter processing. Some authors (see Båth, 1974, pp. 267–269) call this technique *wavenumber–wavenumber filtering*. Six different second-derivative filters have been discussed in detail by Meskó (1966). An approach via Wiener filter theory to construct *optimum second-derivative* and *downward-continuation filters* has been employed by Clarke (1969). See also Section 4.3.3. Smoothing, or low-pass filtering, of two-dimensional observational data has been treated e.g. by Strakhov and Lapina (1967).

In his rather theoretical work, Sax (1966) applied two-dimensional filtering to reduce the ambiguity of observational gravity data. It was required to enhance the effects of density changes within a given layer (signal) whereas effects caused by density changes in all other layers (noise) were to be minimized. The system function of an effective filter could be defined in terms of the wavenumber response at the earth's surface to a density disturbance in the given layer relative to disturbances in other layers.

Sometimes the downward-continuation or second-derivative filters are not best suited for a given problem. If this is the case, filters of other kinds may be more successful. For example Zurflueh (1967) describes two-dimensional wavelength filters for separating gravity and magnetic anomalies of different horizontal extent. Three types of phase-distortionless filters, namely the regional (low-pass), residual (high-pass) and band-pass filters are explained. Since these filters are based merely upon the wavelength separation, they may be applied to any kind of two-dimensional observational data. Lavin and Devane (1970) derived a closed-form solution for constructing two-dimensional low-pass filters. Their filters are phase-distortionless, have a flat pass region and adjustable cut-off wavenumbers. Low-pass wavelength filters are also discussed by Ulrych (1968). Clarke (1971) described the usage of a multiple-input space filter for reducing terrain effects on geophysical maps.

An extension of the theory of one-dimensional Strakhov filters to two-dimensional filters is presented by Naidu (1967). An algorithm for filter coefficients which extracts gravity and magnetic data from a background noise is given. It is supposed that the noise properties are uniquely determined by its autocorrelation function.

6.5.4 *Velocity filtering by means of two-dimensional arrays*

In the case of linear arrays there is no discrimination of varying station-to-source azimuth. Therefore, in Section 6.1 it has been assumed that the source is located on the extended line of sensors. However, two-dimensional arrays with given geometry make it possible to estimate the apparent velocity as well as the azimuth when two or more independent signal move-outs are available. With respect to the array origin, the apparent velocity and horizontal wavenumber may show any direction in the horizontal plane depending upon the station-to-source azimuth. When processing data from plane arrays it is convenient to treat the two quantities as vectors. In this subsection we shall further assume that various frequency filters may be included in individual array channels.

Consider an array with M elements distributed in a certain pattern on the earth's surface. Seismometer positions, with respect to the array origin, are specified by vectors, r_m, or by coordinate pairs, x_m, y_m, for $m = 1, 2, \ldots, M$. The unit-impulse response and complex frequency response of the mth channel are $h_m(t)$ and $H_m(f)$, respectively. Functions $h_m(t)$ and $H_m(f)$ will here include the seismometer as well as all filters present in the mth recording channel. Coherent signal components travel across the array site with their apparent velocities, V, so that, except for vertically propagating signals, time shifts, t_m, appear at individual sensor outputs. The time shift of the mth seismometer related to the array origin (Fig. 6.2a) is given by [6.5]. It is obvious that when the x- and y-axes agree with the W–E and S–N directions, then θ becomes the station-to-source azimuth. Plane wavefronts and constant apparent velocity across the array site are assumed. The response of the mth sensor to a unit impulse propagating across the array with parameters V, θ is:

$$h_m(t - t_m) = h_m\left[t - \frac{r_m \cos(\theta - \alpha_m)}{V}\right] \qquad [6.18]$$

Note that whereas the time-impulse response, $h_m(t)$, is dependent upon time only, $h_m(t - t_m)$ is a function of time and location of the respective sensor. Thus, the processing in terms of $h_m(t - t_m)$ can be visualized as *time-space filtering* (see also Gangi and Disher, 1968). Applying the time-shifting theorem to function [6.18] for the corresponding transform we have:

$$\mathcal{F}\{h_m(t - t_m)\} = H_m(f) \exp[-j2\pi f r_m \cos(\theta - \alpha_m)/V]$$
$$= H_m(f) \exp[-j2\pi r_m k \cos(\theta - \alpha_m)] \qquad [6.19]$$

where $k = 1/\lambda$ is the signal wavenumber and λ is the signal wavelength. Making use of scalar product of two vectors, [6.19] may be simplified as:

$$\{h_m(t - t_m)\} = H_m(f, \mathbf{k}) = H_m(f) \exp(-j2\pi \mathbf{r}_m \cdot \mathbf{k}) \qquad [6.20]$$

where $\mathbf{k}$ is the vector wavenumber. The three-dimensional function [6.20] is the *frequency-wavenumber response function* related to the mth sensor. Due to the linearity property, the response of the complete array, divided by the total number of sensors, becomes:

$$H(f, \mathbf{k}) = \frac{1}{M} \sum_{m=1}^{M} H_m(f, \mathbf{k}) = \frac{1}{M} \sum_{m=1}^{M} H_m(f) \exp(-j2\pi \mathbf{r}_m \cdot \mathbf{k}) \qquad [6.21]$$

which is a system function of a three-dimensional filter realized by means of a plane array with M sensors. Decomposing the vector wavenumber into its x- and y-components, [6.21] may be rewritten as:

$$H(f, k_x, k_y) = \frac{1}{M} \sum_{m=1}^{M} H_m(f) \exp[-j2\pi(x_m k_x + y_m k_y)] \qquad [6.22]$$

A graphical display of the three-dimensional function, $H(f, k_x, k_y)$, is difficult.

VELOCITY FILTERING   151

To surmount this difficulty it is rather common to plot the so-called *k-plane response* for discrete frequencies we are interested in. Since the absolute value of the response [6.22] is of most interest, the amplitude response, $|H(k_x, k_y)|$, or the power response, $\text{Re}^2 H(k_x, k_y) + \text{Im}^2 H(k_x, k_y)$, are most likely to be used. In principle, it is possible to plot the gain or the power response vs. wavenumber for a given azimuth (two-dimensional diagram), see e.g. Capon et al. (1968), or to use a contoured plot above the *k*-plane for all azimuths (three-dimensional diagram), see e.g. Burg (1964). Inserting proper time shifts, t_m, to individual channels, the main lobe of $H(f, k)$ may be transferred to any desired point in the (f, k) space. Consequently, $H(f, k)$ will pass only signals arriving from the expected direction and with apparent velocities within the prescribed range. For example, Hannon and Kovach (1966) used this velocity filtering to separate core phases with apparent velocities ranging from 24 km/sec to 100 km/sec.

Considering signals and coherent noise to be equally likely to arrive from any direction, then the corresponding signal and noise characteristics in the (f, k) space may be diagrammed by making use of conical boundaries obtained by letting the wedges in Fig. 6.4 rotate about the frequency axis. Coherent signals propagating across the array with apparent velocity $V \geq 8$ km/sec are contained within the conical surface $f/|k| = 8$ km/sec, provided that the propagation is nondispersive. Coherent noise with apparent velocities between 2.5 km/sec and 4 km/sec is located between the conical surfaces $f/|k| = 2.5$ km/sec and $f/|k| = 4$ km/sec. Incoherent noise at any particular frequency shows a uniform distribution over the entire *k*-plane. It follows from our model that employing a plane array, complete discrimination between coherent signals and coherent noise is possible while complete elimination of incoherent noise is not.

Velocity filtering may be accomplished by properly shaping the response function $H(f, k_x, k_y)$. In the simplest case, neglect the frequency filters $H_m(f)$ and consider identical wide-band seismometers over the entire array. Then, the response becomes a function of the vector wavenumber only. It is obvious that this approach is in fact the SS technique applied to a two-dimensional array. Since $H(k_x, k_y)$ does not vary with frequency, the straight summation will show reasonably good results only within a limited range of frequencies (Burg, 1964).

Further improvement may be achieved by applying the FS method. This requires an appropriately designed three-dimensional filtering technique which is applied prior to the summation. In [6.22] frequency filters together with seismometers are absorbed in the $H_m(f)$ functions. Note that $H_m(f)$ depends upon frequency and location within the array pattern. Due to the finite spatial sampling, the design of $H_m(f)$ leads to multi-channel processing. Among methods that specify the filter $H_m(f)$ to fulfill given requirements, the *multi-channel Wiener filtering* and *maximum-likelihood methods* are probably the most frequently used in seismic applications. The maximum-likelihood method includes frequency filtering of individual array channels but provides minimum frequency distortion of the output signal. The sum of all M filters provides wide-band flat frequency characteristics in spite of the fact that individual filters may show quite different behaviour. Thus, waveforms arriving from an expected direction with expected apparent velocity are passed more or less undistorted while all other waveforms are suppressed. As discussed before, the Wiener filtering minimizes the mean-square deviation between the output and the desired signal. In general, the two methods form an array output waveform which may be used as an estimate of the unknown arriving signal. Details are discussed by Levin (1964), Capon and Greenfield (1965), Green et al. (1966), Farrell (1971) and others.

As an example, we may mention results obtained via the multi-channel Wiener filtering technique. Burg (1964) shows that in order to perform Wiener-sense optimum filtering for given M and array geometry, the filters $H_m(f)$ must satisfy the conditions:

$$\sum_{n=1}^{M} [S_{mn}(f) + N_{mn}(f)] H_n^*(f) = S_{mo}(f) \quad \text{for } m = 1, 2, \ldots, M$$

where $S_{mn}(f)$ is the cross-power spectrum between signals at the mth and nth seismometer, $N_{mn}(f)$ is the related noise spectrum and $S_{mo}(f)$ is the cross-power spectrum between signals at the mth seismometer and the array origin. The asterisk superscript indicates the complex-conjugate function. Numerical results presented by Burg illustrate the superiority of the multi-channel Wiener filtering when compared with the straight summation or single-channel Wiener filtering applied to the summed array output. Corresponding time-domain solution has been discussed by Hurbal (1972). Laster and Linville (1966) present a detailed discussion on employment of optimum multi-channel filters for separation of dispersive modes.

Five different multi-channel filters designed according to different signal and noise models have been described by Backus et al. (1964). Performance of these filters has been tested for a 19-element array at Cumberland Plateau Seismological Observatory. In general, multi-channel filtering shows substantial improvement of the signal/noise ratio when compared with the SS technique. Schneider et al. (1965) employed a multi-channel filter for efficient rejection of multiple reflections. They combine the concept of horizontal common-depth-point stacking technique with a Wiener-sense optimum three-channel filter described by Burg (1964).

At the present time the great majority of operating arrays employ sensors located at or close to the earth's surface. However, theoretical calculations together with observational data reveal that the noise level decreases with increasing depth. Thus, an array consisting of buried sensors should produce a further improvement of the signal/noise ratio. Roden (1965, 1968) carried out theoretical investigations and showed that an array with shallow-buried seismometers together with a multi-channel filter is superior to surface array with otherwise equal parameters. On the other hand, outputs from tested surface and *vertical arrays* do not indicate any significant difference. When vertical arrays are used, the effect of interference between direct and surface-reflected waves creates an additional problem for the processing scheme. Another obstacle in utilizing vertical arrays is that if we move the seismometer farther away from the free surface not only noise but also signal levels decrease. By the word "shallow" Roden means a depth of about 100 m or more (a few hundred feet or more) where the level of the incoherent noise is expected to be already sufficiently low when compared with the signal level.

Der (1970) describes an experiment with a vertical array equipped with three-component sensors. Performances of horizontal and vertical arrays at sea have been compared by Merzer (1972).

REFERENCES

Note. Due to the different transliteration systems used Yanovskiĭ = Yanovskiy.

Ackroyd, M. H., 1973. *Digital Filters*. Butterworths, 82 pp.
Agarwal, B. N. P. and Lal, T., 1970. Application of frequency analysis in two-dimensional gravity interpretation. *Geoexploration*, 10: 91–100.
Aguilera, R., De Bremaecker, J. C. and Hernandez, S., 1970. Design of recursive filters. *Geophysics*, 35: 247–253.
Allsopp, D. F., Burke, M. D. and Cumming, G. L., 1972. A digital seismic recording system. *Bull. Seismol. Soc. Am.*, 62: 1641–1648.
An, V. A., 1965. On the possibility of using analog–digital conversion for recording microvariations in the earth's electromagnetic field. *Izv. Acad. Sci. USSR, Phys. Solid Earth*, 9: 647–649 (Engl. ed.).
Anders, E. B., Johnson, J. J., Lasaine, A. D., Spikes, P. W. and Taylo, J. T., 1964. *Digital Filters*. NASA Contr. Rep. CR-136, 132 pp.
Anstey, N. A., 1964. Correlation techniques – a review. *Geophys. Prospect.*, 12: 355–382.
Backus, M. M., 1959. Water reverberations – their nature and elimination. *Geophysics*, 24: 233–261.
Backus, M., Burg, J., Baldwin, D. and Bryan, E., 1964. Wide-band extraction of mantle P-waves from ambient noise. *Geophysics*, 29: 672–692.
Båth, M., 1968. *Mathematical Aspects of Seismology*. Elsevier, 415 pp.
Båth, M., 1974. *Spectral Analysis in Geophysics*. Elsevier, 563 pp.
Bayless, J. W. and Brigham, E. O., 1970. Application of the Kalman filter to continuous signal restoration. *Geophysics*, 35: 2–23.
Belotelov, V. L. and Rykunov, L. N., 1963. A device for digitizing seismograms. *Izv. Acad. Sci. USSR, Geophys. Ser.*, 3: 293–294 (Engl. ed.).
Bendat, J. S. and Piersol, A. G., 1971. *Random Data: Analysis and Measurement Procedures*. Wiley-Interscience, 407 pp.
Berckhemer, H. and Jacob, K. H., 1968. Investigation of the dynamical process in earthquake foci by analyzing the pulse shape of body waves. *Ber. Inst. Meteorol. Geophys. Univ. Frankfurt/Main*, 13: 1–85.
Bhattacharyya, B. K., 1972. Design of spatial filters and their application to high-resolution aeromagnetic data. *Geophysics*, 37: 68–91.
Bhimasankaram, V. L. S., Tarkhov, A. G., Nikitin, A. A. and Seshagiri Rao, S. V., 1973. Interprofile correlation and self-setting filtration method of analysis of geophysical data. *Geophys. Prospect.*, 21: 464–471.
Birtill, J. W. and Whiteway, F. E., 1965. The application of phased arrays to the analysis of seismic body waves. *Philos. Trans. R. Soc. London, Ser. A*, 258: 421–493.
Black, D. I. and Scollar, I., 1969. Spatial filtering in the wave-vector domain. *Geophysics*, 34: 916–923.
Blackman, R. B., 1965. *Linear Data-Smoothing and Prediction in Theory and Practice*. Addison-Wesley, 182 pp.
Blackman, R. B. and Tukey, J. W., 1958. *The Measurement of Power Spectra*. Dover Publications, 190 pp.
Bogert, B. P., 1962. Correction of seismograms for the transfer function of the seismometer. *Bull. Seismol. Soc. Am.*, 52: 781–792.

Botezatu, R. and Calota, C., 1973. Cross-correlation as an aid in simultaneous gravity and magnetic analysis. *Geophys. Prospect.*, 21: 472–483.
Boyarskiy, E. A. and Kogan, M. G., 1968. Linear smoothing of measurements made with marine gravimeters. *Izv. Acad. Sci. USSR, Phys. Solid Earth*, 10: 640–643 (Engl. ed.).
Bracewell, R., 1965. *The Fourier Transform and Its Applications.* McGraw-Hill, 381 pp.
Brigham, E. O., Smith, H. W., Jr., Bostick, F. X., Jr. and Duesterhoeft, W. C., Jr., 1968. An iterative technique for determining inverse filters. *IEEE Trans.*, GE-6: 86–96.
Bruland, L. and Rygg, E., 1971. Experiments with chirp filtering of surface waves. *Rep. No. 6, Seismol. Obs., Univ. of Bergen*, 34 pp.
Buben, J. and Rudajev, Vl., 1974. Statistical prediction of rock bursts. (In Czech with an English summary) *Publ. Inst. Geophys. Pol. Acad. Sci.*, 67: 3–31.
Burch, J. J., Green, A. W., Jr. and Grote, H. H., 1964. Restoration and correction of time functions by the synthesis of inverse filters on analog computers. *IEEE Trans.*, GE-2: 19–24.
Burg, J. P., 1964. Three-dimensional filtering with an array of seismometers. *Geophysics*, 29: 693–713.
Burke, M. D., Kanasewich, E. R., Malinsky, J. D. and Montalbetti, J. F., 1970. A wide-band digital seismograph system. *Bull. Seismol. Soc. Am.*, 60: 1417–1426.
Burns, W. R., 1968. A statistically optimized deconvolution. *Geophysics*, 33: 255–263.
Byerly, P. E., 1965. Convolution filtering of gravity and magnetic maps. *Geophysics*, 30: 281–283.
Capon, J. and Green, P. E., 1968. Recent results from the Large-Aperture Seismic Array. *Supplemento al Nuovo Cimento*, 6: 82–95.
Capon, J. and Greenfield, R. J., 1965. Asymptotically optimum multidimensional filtering for sampled-data processing of seismic arrays. *M.I.T. Lincoln Lab., Tech. Note* 1965-57, 39 pp.
Capon, J., Greenfield, R. J., Kolker, R. J. and Lacoss, R. T., 1968. Short-period signal processing results for the Large Aperture Seismic Array. *Geophysics*, 33: 452–472.
Capon, J., Greenfield, R. J. and Lacoss, R. T., 1969. Long-period signal processing results for the Large Aperture Seismic Array. *Geophysics*, 34: 305–329.
Cassano, E. and Rocca, F., 1974. After-stack multichannel filters without mixing effects. *Geophys. Prospect.*, 22: 330–344.
Chan, S. H. and Leong, L. S., 1972. Analysis of least-squares smoothing operators in the frequency domain. *Geophys. Prospect.*, 20: 892–900.
Chan, S. H. and Leong, L. S., 1974. Filtering of discrete time series by symmetric least-squares operators. *Math. Geol.*, 6: 153–171.
Choy, G. and McCamy, K., 1973. Enhancement of long-period signals by time-varying adaptive filters. *J. Geophys. Res.*, 78: 3505–3511.
Claerbout, J. F., 1964. Detection of P-waves from weak sources at great distances. *Geophysics*, 29: 197–211.
Clarke, G. K. C., 1968. Time-varying deconvolution filters. *Geophysics*, 33: 936–944.
Clarke, G. K. C., 1969. Optimum second-derivative and downward-continuation filters. *Geophysics*, 34: 424–437.
Clarke, G. K. C., 1971. Linear filters to suppress terrain effects on geophysical maps. *Geophysics*, 36: 963–966.
Clay, C. S. and Liang, W. L., 1962. Continuous seismic profiling with matched filter detector. *Geophysics*, 27: 786–795.
Clement, W. G., 1973. Basic principles of two-dimensional digital filtering. *Geophys. Prospect.*, 21: 125–145.
Cochran, M. D., 1973. Seismic signal detection using sign bits. *Geophysics*, 38: 1042–1052.
Crampin, S. and Båth, M., 1965. Higher modes of seismic surface waves: Mode separation. *Geophys. J. R. Astron. Soc.*, 10: 81–92.

Crump, N. D., 1974. A Kalman filter approach to the deconvolution of seismic signals. *Geophysics*, 39: 1–13.

Dampney, C. N. G., 1965. Three criteria for the judgement of vertical continuation and derivative methods of geophysical interpretation. *Geoexploration*, 4: 3–24.

Darby, E. K. and Davies, E. B., 1967. The analysis and design of two-dimensional filters for two-dimensional data. *Geophys. Prospect.*, 15: 383–406.

Davenport, W. B., Jr. and Root, W. L., 1958. *An Introduction to the Theory of Random Signals and Noise*. McGraw-Hill, 393 pp.

Davies, E. B. and Mercado, E. J., 1968. Multichannel deconvolution filtering of field recorded seismic data. *Geophysics*, 33: 711–722.

Dean, W. C., 1965. P-wave correlations and array alignments. *Proc. IEEE*, 53: 1861–1865.

De Bremaecker, J. C., Donoho, P. and Michel, J. G., 1962. A direct digitizing seismograph. *Bull. Seismol. Soc. Am.*, 52: 661–672.

De Bremaecker, J. C., Sitton, G. A., Rusk, S. K., Graham, M. H. and Schutz, T. C., 1963. The Rice digital seismograph system. *J. Geophys. Res.*, 68: 5029–5034.

Del Toro, V. and Parker, S. R., 1960. *Principles of Control Systems Engineering*, McGraw-Hill, 686 pp.

Denham, D., 1963. The use of geophone groups to improve the signal-to-noise ratio of the first arrival in refraction shooting. *Geophys. Prospect.*, 11: 389–408.

Der, Z. A., 1970. Some data processing results for a vertical array of triaxial seismometers. *Geophysics*, 35: 337–343.

d'Hoeraene, J., 1962. Déconvolution de traces réelles. *Geophys. Prospect.*, 10: 68–83.

Dobrin, M. B. and Ward, S. H., 1962. Tools for tomorrow's geophysics. *Geophys. Prospect.*, 10: 433–452.

Domenico, S. N., 1965. Phase-distortionless filtering. *Geophysics*, 30: 32–50.

Donnell, W. F., 1967. Sources of error in a seismic digital recording system. *Geophys. Prospect.*, 15: 246–261.

Embree, P., Burg, J. P. and Backus, M. M., 1963. Wide-band velocity filtering — the pie-slice process. *Geophysics*, 28: 948–974.

Fail, J. P. and Brau, G., 1963. Les filtres en éventail. *Geophys. Prospect.*, 11: 131–163.

Fail, J. P. and Layotte, P. C., 1970. Méthode de filtrage du fantôme: Application à des cas réels. *Geophys. Prospect.*, 28: 434–464.

Farrell, E. J., 1971. Sensor-array processing with channel-recursive bayes technique. *Geophysics*, 36: 822–834.

Finetti, I., Nicolich, R. and Sancin, S., 1971. Review on the basic theoretical assumptions in seismic digital filtering. *Geophys. Prospect.*, 19: 292–320.

Flinn, E. A., 1965. Signal analysis using rectilinearity and direction of particle motion. *Proc. IEEE*, 53: 1874–1876.

Floyd, G. F., 1969. Gravimeter filters. *Geophysics*, 34: 968–973.

Ford, W. T., 1967. Application of Green's theorem in two-dimensional filtering. *Geophysics*, 32: 739–740.

Ford, W. T. and Hearne, J. H., 1966. Least-squares inverse filtering. *Geophysics*, 31: 917–926.

Foster, M. R., Sengbush, R. L. and Watson, R. J., 1964. Design of sub-optimum filter systems for multi-trace seismic data processing. *Geophys. Prospect.*, 12: 173–191.

Foster, M. R., Sengbush, R. L. and Watson, R. J., 1968. Use of Monte Carlo techniques in optimum design of the deconvolution process. *Geophysics*, 33: 945–949.

Frank, H. R. and Doty, W. E. N., 1953. Signal-to-noise ratio improvements by filtering and mixing. *Geophysics*, 18: 587–604.

Frasier, C. W., 1972. Observations of pP in the short-period phases of NTS explosions recorded at Norway. *Geophys. J. R. Astron. Soc.*, 31: 99–109.

REFERENCES

Galbraith, J. N., 1971. Prediction error as a criterion for operator length. *Geophysics*, 36: 261-265.
Galbraith, J. N., Jr. and Wiggins, R. A., 1968. Characteristics of optimum multichannel stacking filters. *Geophysics.*, 33: 36-48.
Galli, M. and Randi, P., 1967. On the design of the optimum numerical filter with a prefixed response. *Ann. Geofis.*, 20: 401-414.
Gangi, A. F. and Disher, D., 1968. A space-time filter for seismic models. *Geophysics*, 33: 88-104.
George, C. F., Jr., Smith, H. W. and Bostick, F. X., Jr., 1964. Application of inverse filters to induction log analysis. *Geophysics*, 29: 93-104.
Gjöystdal, H. and Husebye, E. S., 1972. A comparison of performance between prediction error and bandpass filters. *NORSAR Tech. Rep.*, No. 48, 8 pp.
Gold, B. and Rader, C. M., 1969, *Digital Processing of Signals*. McGraw-Hill, 269 pp.
Göncz, G. and Zelei, A., 1972. Recursion band-filters and their design. *Geophys. Trans. Hung. Geophys. Inst. Roland Eötvös*, 3-4: 59-71.
Green, P. E., Jr., 1965. A Large Aperture Seismic Array. *M.I.T. Lincoln Lab., Group Rep.* 1965-1, 25 pp.
Green, P. E., Jr., Frosch, R. A. and Romney, C. F., 1965. Principles of an experimental Large Aperture Seismic Array (LASA). *Proc. IEEE*, 53: 1821-1833.
Green, P. E., Jr., Kelly, E. J., Jr., and Levin, M. J., 1966. A comparison of seismic array processing methods. *Geophys. J. R. Astron. Soc.*, 11: 67-84.
Gunn, P. J., 1972. Application of Wiener filters to transformations of gravity and magnetic fields. *Geophys. Prospect.*, 20: 860-871.
Hales, F. W. and Edwards, T. E., 1955. Some theoretical considerations on the use of multiple geophones arranged linearly along the line of traverse. *Geophys. Prospect.*, 3: 65-73.
Hamming, R. W., 1962. *Numerical Methods for Scientists and Engineers*. McGraw-Hill, 411 pp.
Hannon, W. J. and Kovach, R. L., 1966. Velocity filtering of seismic core phases. *Bull. Seismol. Soc. Am.*, 56: 441-454.
Hartenberger, R. A. and Van Nostrand, R. G., 1972. Influence of number and spacing of sensors on the effectiveness of seismic arrays. *Geophys. Prospect.*, 20: 771-784.
Haubrich, R. A., 1968. Array design. *Bull. Seismol. Soc. Am.*, 58: 977-991.
Haubrich, R. A. and Iyer, H. M., 1962. A digital seismograph system for measuring earth noise. *Bull. Seismol. Soc. Am.*, 52: 87-93.
Helstrom, C. W., 1960. *Statistical Theory of Signal Detection*. Pergamon Press, 364 pp.
Herrmann, R. B., 1973. Some aspects of band-pass filtering of surface waves. *Bull. Seismol. Soc. Am.*, 63: 663-672.
Holloway, J. L., Jr., 1958. Smoothing and filtering of time series and space fields. *Adv. Geophys.*, 4: 351-389.
Holtz, H. and Leondes, C. T., 1966. The synthesis of recursive digital filters. *J. Assoc. Comput. Mach.*, 13: 262-280.
Holzman, M., 1963. Chebyshev optimized geophone arrays. *Geophysics*, 28: 145-155.
Howell, B. F., Jr., Lavin, P. M., Watson, R. J., Cheng, Y. Y. and Lin, J. L., 1967. Method for recognizing repeated pulse sequences in a seismogram. *J. Geophys. Res.*, 72: 3225-3232.
Huang, Y. T., 1966. Spectral analysis of digitized seismic data. *Bull. Seismol. Soc. Am.*, 56: 425-440.
Hurbal, P., 1972. Three-dimensional optimum multichannel velocity filters. *Geophys. Prospect.*, 20 : 28-46.
Iyer, H. M., 1968. Determination of frequency-wave-number spectra using seismic arrays. *Geophys. J. R. Astron. Soc.*, 16: 97-117.
Jackson, P. L., 1967. Truncation and phase relationships of sinusoids. *J. Geophys. Res.*, 72: 1400-1403.
Jones, H. J. and Morrison, J. A., 1954. Cross-correlation filtering. *Geophysics*, 19: 660-683.

Jones, H. J., Morrison, J. A., Sarrafian, G. P. and Spieker, L. J., 1955. Magnetic delay line filtering techniques. *Geophysics*, 20: 745–765.
Jury, E. I., 1958. *Sampled-Data Control Systems*. Wiley, 453 pp.
Jury, E. I., 1964. *Theory and Application of the z-Transform Method*. Wiley, 330 pp.
Kaiser, J. F., 1966. Digital filters. In: F. F. Kuo and J. F. Kaiser (Editors), *System Analysis by Digital Computers*. Wiley, 438 pp.
Kanasewich, E. R., 1973. *Time Sequence Analysis in Geophysics*. The University of Alberta Press, 352 pp.
Kanasewich, E. R., Hemmings, C. D. and Alpaslan, T., 1973. Nth-root stack nonlinear multichannel filter. *Geophysics*, 38: 327–338.
Kats, S. A., 1972. Discrete quasi-optimal and minimax filters. *Izv. Acad. Sci. USSR, Phys. Solid Earth*, 9: 594–598 (Engl. ed.).
King, D. W., Mereu, R. F. and Muirhead, K. J., 1973. The measurement of apparent velocity and azimuth using adaptive processing techniques on data from the Warramunga seismic array. *Geophys. J. R. Astron. Soc.*, 35: 137–167.
Klíma, K. and Kulhánek, O., 1970. Seismic signals processing by using the method of digital deconvolution. *Geophys. J. R. Astron. Soc.*, 21: 403 (abstract).
Kondrat'yev, I. K., 1968. On the theory of inverse digital filtering of wave processes. *Izv. Acad. Sci. USSR, Phys. Solid Earth*, 4: 231–237 (Engl. ed.).
Koopmans, L. H., 1961. An evaluation of a signal-summing technique for improving the signal-to-noise ratio for seismic events. *J. Geophys. Res.*, 66: 3879–3893.
Koopmans, L. H., 1974. *The Spectral Analysis of Time Series*. Academic Press, 366 pp.
Krey, T. and Toth, F., 1973. Remarks on wavenumber filtering in the field. *Geophysics*, 38: 959–970.
Ku, C. C., Telford, W. M. and Lim, S. H., 1971. The use of linear filtering in gravity problems. *Geophysics*, 36: 1174–1203.
Kulhánek, O., 1967. Seismic noise filtering using digital computers. *Trav. Inst. Geophys. Acad. Tchécoslov. Sci.*, 273: 255–286.
Kulhánek, O., 1973. Signal and noise coherence determination for the Uppsala Seismograph Array Station. *Pure Appl. Geophys.*, 109: 1653–1671.
Kulhánek, O. and Klíma, K., 1970. The reliable frequency band for amplitude spectra corrections. *Geophys. J. R. Astron. Soc.* 21: 235–242.
Kunetz, G. and Fourmann, J. M., 1968. Efficient deconvolution of marine seismic records. *Geophysics*, 33: 412–423.
Kurita, T., 1969. Spectral analysis of seismic waves. Part 1. Data windows for the analysis of transient waves. *Spec. Contrib. Geophys. Inst., Kyoto Univ.*, 9: 97–122.
Lacoss, R. T., 1965. Geometry and patterns of Large Aperture Seismic Arrays. *M.I.T. Lincoln Lab., Tech. Note* 1965-64, 83 pp.
Landisman, M., Dziewonski, A. and Satô, Y., 1969. Recent improvements in the analysis of surface wave observations. *Geophys. J. R. Astron. Soc.*, 17: 369–403.
Lapina, M. I. and Strakhov, V. N., 1973. Formalized algorithms for the filtration of potential fields. *Izv. Acad. Sci. USSR, Geophys. Ser.*, 9: 442–453 (Engl. ed.).
Laster, S. J. and Linville, A. F., 1966. Application of multichannel filtering to the separation of dispersive modes of propagation. *J. Geophys. Res.*, 71: 1669–1701.
Lavin, P. M. and Devane, J. F., 1970. Direct design of two-dimensional digital wavenumber filters. *Geophysics*, 35: 1073–1078.
Lee, Y. W., 1960. *Statistical Theory of Communication*. Wiley, 509 pp.
Levin, M. J., 1964. Maximum-likelihood array processing. *Seismic Discrimination, M.I.T. Lincoln Lab., Semiann. Tech. Summ.*, 31: 21–23.
Levin, M. J. and Price, R., 1964. Automatic alarm systems. *Seismic Discrimination, M.I.T. Lincoln Lab., Semiann. Tech. Summ.*, 30: 15–17.

Levinson, N., 1949. *The Wiener RMS (root mean square) error criterion in filter design and prediction*. Appendix in: Wiener (1949, pp. 129–148).
Lindsey, J. P., 1960. Elimination of seismic ghost reflections by means of a linear filter. *Geophysics*, 25: 130–140.
Lombardi, L. V., 1955. Notes on the use of multiple geophones. *Geophysics*, 20: 215–226.
Longman, I. M. and Sharir, M., 1971. Laplace transform inversion of rational functions. *Geophys. J. R. Astron. Soc.*, 25: 299–305.
Manzoni, G., 1967. Theoretical evaluation of the perturbation on power spectra due to random errors in the spacing of the sampling instants. *Boll. Geofis. Teor. Appl.*, 9: 248–252.
McClellan, J. H. and Parks, T. W., 1972. Equiripple approximation of fan filters. *Geophysics*, 37: 573–583.
Melton, B. S., 1967. Analog-to-digital conversion—a problem or "decibels to digits." *IEEE Trans.*, GE-5: 18–25.
Mercado, E. J., 1968. Linear phase filtering of multicomponent seismic data. *Geophysics*, 33: 926–935.
Merkel, R. H. and Alexander, S. S., 1969. Use of correlation analysis to interpret continental margin ECOOE refraction data. *J. Geophys. Res.*, 74: 2683–2697.
Merzer, A. M., 1972. Horizontal and vertical arrays at sea. *Geophys. J. R. Astron. Soc.*, 29: 367–370.
Meskó, A., 1965. Some notes concerning the frequency analysis for gravity interpretation. *Geophys. Prospect.*, 13: 475–488.
Meskó, C. A., 1966. Two-dimensional filtering and the second derivative method. *Geophysics*, 31: 606–617.
Meyerhoff, H. J., 1966. A self-adjusting filter for shallow seismic exploration. *Geophysics*, 31: 340–345.
Meyerhoff, H. J., 1968a. Realization of sharp cut-off frequency characteristics on digital computers (Part I). *Geophys. Prospect.*, 16: 208–219.
Meyerhoff, H. J., 1968b. Realization of sharp cut-off frequency characteristics on digital computers (Part II). *Geophys. Prospect.*, 16: 220–246.
Meyerhoff, H. J., 1968c. Realization of sharp cut-off frequency characteristics on digital computers (Part III). *Geophys. Prospect.*, 16: 491–510.
Mikulski, Z. and Mikulska, M., 1973. Untersuchung der Periodizität von hydrometeorologischen Erscheinungen nach der Autokorrelationsmethode von Fuhrich. *Gerlands Beitr. Geophys.*, 82: 187–193.
Miller, W. F., 1963. The Caltech digital seismograph. *J. Geophys. Res.*, 68: 841–847.
Moltshan, G. M., Pissarenko, V. F. and Smirnova, N. A., 1964. Some statistical methods of detecting signals in noise. *Geophys. J. R. Astron. Soc.*, 8: 319–323.
Montalbetti, J. F. and Kanasewich, E. R., 1970. Enhancement of teleseismic body phases with a polarization filter. *Geophys. J. R. Astron. Soc.*, 21: 119–129.
Mooney, H. M., 1968. Pole-and-zero design of digital filters. *Geophysics*, 33: 354–360.
Mufti, I. R., 1972. Design of small operators for the continuation of potential field data. *Geophysics*, 37: 488–506.
Muir, F. and Hales, F. W., 1955. A rational approach to the design of electrical filters and of shothole and geophone patterns in seismic reflection prospecting. *Geophys. Prospect.*, 3: 350–358.
Naidu, P. S., 1967. Two dimensional Strakhov's filter for extraction of potential field signal. *Geophys. Prospect.*, 15: 135–150.
Naidu, P. S., 1968. An example of linear filtering in aeromgnetic interpretation. *Geophysics*, 33: 602–612.
Nakhamkin, S. A., 1969. Fan filtration. *Izv. Acad. Sci. USSR, Phys. Solid Earth*, 11: 686–691 (Engl. ed.).

Neunhöfer, H., 1971. Deconvolution of the seismogram concerning the parameters of the seismograph. *Gerlands Beitr. Geophys.*, 80: 475–482.

Nikitin, A. A. and Yanovskiy, A. K., 1973. Algorithms for recursive filtering for the numerical evaluation of seismic recordings. *Izv. Acad. Sci. USSR, Phys. Solid Earth*, 4: 227–232 (Engl. ed.).

Ormsby, J. F. A., 1961. Design of numerical filters with applications to missile data processing. *J. Assoc. Comput. Mach.*, 8: 440–466.

Otnes, R. K. and Enochson, L., 1972. *Digital Times Series Analysis*. Wiley, 467 pp.

Ott, N. and Meder, H. G., 1972. The Kalman filter as a prediction error filter. *Geophys. Prospect.*, 20: 549–560.

Papoulis, A., 1962. *The Fourier Integral and its Applications*. McGraw-Hill, 318 pp.

Peacock, K. L. and Treitel, S., 1969. Predictive deconvolution: Theory and practice. *Geophysics*, 34: 155–169.

Petrov, L. V., 1963. The use of Chebyshev polynomials in the design of arrays for seismic exploration. *Izv. Acad. Sci. USSR, Geophys. Ser.*, 5: 431–440 (Engl. ed.).

Phinney, R. A. and Smith, S. W., 1963. Processing of seismic data from an automatic digital recorder. *Bull. Seismol. Soc. Am.*, 53: 549–562.

Rabiner, L. R., Cooley, J. W., Helms, H. D., Jackson, L. B., Kaiser, J. F., Rader, C. M., Schafer, R. W., Steiglitz, K. and Weinstein, C. J., 1972. Terminology in digital signal processing. *IEEE Trans.*, AU-20: 322–337.

Rader, C. M. and Gold, B., 1967. Digital filter design techniques in the frequency domain. *Proc. IEEE*, 55: 149–171.

Ragazzini, J. R. and Franklin, G. F., 1958. *Sampled-Data Control Systems*. McGraw-Hill, 331 pp.

Rice, R. B., 1962. Inverse convolution filters. *Geophysics*, 27: 4–18.

Ricker, N., 1940. The form and nature of seismic waves and the structure of seismograms. *Geophysics*, 5: 348–366.

Robinson, E. A., 1957. Predictive decomposition of seismic traces. *Geophysics*, 22: 767–778.

Robinson, E. A., 1967a. Predictive decomposition of time series with application to seismic exploration. *Geophysics*, 32: 418–484.

Robinson, E. A., 1967b. *Statistical Communication and Detection with special reference to Digital Data Processing of Radar and Seismic Signals*. Griffin, 362 pp.

Robinson, E. A. and Treitel, S., 1964. Principles of digital filtering. *Geophysics*, 29: 395–404.

Robinson, E. A. and Treitel, S., 1965. Dispersive digital filters. *Rev. Geophys.*, 3: 433–461.

Robinson, E. A. and Treitel, S., 1967. Principles of digital Wiener filtering. *Geophys. Prospect.*, 15: 311–333.

Robinson, J. C., 1972. Computer-designed Wiener filters for seismic data. *Geophysics*, 37: 235–259.

Roden, R. B., 1965. Horizontal and vertical arrays for teleseismic signal enhancement. *Geophysics*, 30: 597–608.

Roden, R. B., 1968. Seismic experiments with vertical arrays. *Geophysics*, 33: 270–284.

Ryall, A., 1964. Improvement of array seismic recordings by digital processing. *Bull. Seismol. Soc. Am.*, 54: 277–294.

Sakrison, D. J., Ford, W. T. and Hearne, J. H., 1967. The z-transform of a realizable time function. *IEEE Trans.*, GE-5: 33–41.

Savarenskiĭ, E. F. and Kosarev, G. L., 1967. Digital filtration on the long-period vibrations by the 1964 Alaska earthquake. *Izv. Acad. Sci. USSR, Phys. Solid Earth*, 12: 809–810 (Engl. ed.).

Savit, C. H., Brustad, J. T. and Sider, J., 1958. The moveout filter. *Geophysics*, 23: 1–25.

Sax, R. L., 1966. Application of filter theory and information theory to the interpretation of gravity measurements. *Geophysics*, 31: 570–575.

Schneider, W. A., Prince, E. R., Jr. and Giles, B. F., 1965. A new data-processing technique for multiple attenuation exploiting differential normal moveout. *Geophysics*, 30: 348–362.

REFERENCES

Schoenberger, M., 1970. Optimization and implementation of marine seismic arrays. *Geophysics*, 35: 1038–1053.
Sengbush, R. L. and Foster, M. R., 1968. Optimum multichannel velocity filters. *Geophysics*, 33: 11–35.
Sengbush, R. L., Lawrence, P. L. and McDonal, F. J., 1961. Interpretation of synthetic seismograms. *Geophysics*, 26: 138–157.
Shanks, J. L., 1967. Recursive filters for digital processing. *Geophysics*, 32: 33–51.
Shapiro, R., 1970. Smoothing, filtering and boundary effects. *Rev. Geophys.*, 8: 359–387.
Shaub, Yu. B., 1963. The use of correlation analysis for the evaluation of geophysical data. *Izv. Acad. Sci. USSR, Geophys. Ser.*, 4: 358–364 (Engl. ed.).
Shimshoni, M. and Smith, S. W., 1964. Seismic signal enhancement with three-component detectors. *Geophysics*, 29: 664–671.
Shtemenko, Yu. N., 1971. Discrimination of seismic signals with the aid of a self-adaptive decorrelation filter. *Izv. Acad. Sci. USSR, Phys. Solid Earth*, 3: 217–221 (Engl. ed.).
Silverman, D., 1967. The digital processing of seismic data. *Geophysics*, 32: 988–1002.
Simpson, S. M., Jr., 1955. Similarity of output traces as a seismic operator criterion. *Geophysics*, 20: 254–269.
Smith, M. K., 1956. Noise analysis and multiple seismometer theory. *Geophysics*, 21: 337–360.
Smith, M. K., 1958. A review of methods of filtering seismic data. *Geophysics*, 23: 44–57.
Smith, S. W., 1965. Seismic digital data acquisition system. *Rev. Geophys.*, 3: 151–156.
Solodovnikov, V. V., 1960. *Introduction to the Statistical Dynamics of Automatic Control Systems*. Dover Publications, 307 pp.
Somers, H. and Manchee, E. B., 1966. Selectivity of the Yellowknife seismic array. *Geophys. J. R. Astron. Soc.*, 10: 401–412.
Strakhov, V. N. and Lapina, M. I., 1967. A method of smoothing of potential fields. *Izv. Acad. Sci. USSR, Phys. Solid Earth*, 8: 511–520.
Swartz, C. A. and Sokoloff, V. M., 1954. Filtering associated with selective sampling of geophysical data. *Geophysics*, 19: 402–419.
Tolstoy, I., 1973. *Wave Propagation*. McGraw-Hill, 466 pp.
Toman, K., 1965. The spectral shifts of truncated sinusoids. *J. Geophys. Res.*, 70: 1749–1750.
Tou, J. T., 1959. *Digital and Sampled-Data Control Systems*. McGraw-Hill, 631 pp.
Treitel, S., 1970. Principles of digital multichannel filtering. *Geophysics*, 35: 785–811.
Treitel, S. and Robinson, E. A., 1964. The stability of digital filters. *IEEE Trans.*, GE-2: 6–18.
Treitel, S. and Robinson, E. A., 1966a. Seismic wave propagation in layered media in terms of communication theory. *Geophysics*, 31: 17–32.
Treitel, S. and Robinson, E. A., 1966b. The design of high-resolution digital filters. *IEEE Trans.*, GE-4: 25–38.
Treitel, S. and Robinson, E. A., 1969. Optimum digital filters for signal to noise ratio enhancement. *Geophys. Prospect.*, 17: 248–293.
Treitel, S., Shanks, J. L. and Frasier, C. W., 1967. Some aspects of fan filtering. *Geophysics*, 32: 789–800.
Turin, G. L., 1960. An introduction to matched filters. *IRE Trans.*, IT-6: 311–329.
Ulrych, T. J., 1968. Effects of wavelength filtering on the shape of the residual anomaly. *Geophysics*, 33: 1015–1018.
Ulrych, T. J., 1972. Maximum entropy power spectrum of truncated sinusoids. *J. Geophys. Res.*, 77: 1396–1400.
Verma, R. K. and Roy, A., 1970. A graphical method for computing geophone group response. *Geophysics*, 35: 704–707.
Vích, R., 1968. Selective properties of digital filters obtained by convolution approximation. *Electronics Letters*, 4.

Vinnik, L. P., 1963. The space-time filtration of seismic signals. *Izv. Acad. Sci. USSR, Geophys. Ser.*, 6: 521–527 (Engl. ed.).

Wadsworth, G. P., Robinson, E. A., Bryan, J. G. and Hurley, P. M., 1953. Detection of reflections on seismic records by linear operators. *Geophysics*, 18: 539–586.

Wang, R. J., 1969. The determination of optimum gate lengths for time-varying Wiener filtering. *Geophysics*, 34: 683–695.

Wang, R. J. and Treitel, S., 1973. The determination of digital Wiener filters by means of gradient methods. *Geophysics*, 38: 310–326.

White, R. E. and Mereu, R. F., 1972. Deconvolution of refraction seismograms from large underwater explosions. *Geophysics*, 37: 431–444.

Wickens, A. J. and Kollar, F., 1967. A wide range seismogram digitizer. *Bull. Seismol. Soc. Am.*, 57: 91–98.

Wiener, N., 1949. *Extrapolation, Interpolation, and Smoothing of Stationary Time Series with Engineering Applications.* The Technology Press of M.I.T. and Wiley, 163 pp.

Wiggins, R. A., 1966. ω-k filter design. *Geophys. Prospect.*, 14: 427–440.

Wiggins, R. A., 1967. Use of expected error in the design of least-squares optimum filters. *Geophys. Prospect.*, 15: 288–296.

Wood, L. C., 1968. A review of digital pass filtering. *Rev. Geophys.*, 6: 73–97.

Wood, L. C. and Hockens, S. N., 1970. Least-squares smoothing operators. *Geophysics*, 35: 1005–1019.

Yanovskiĭ, A. K., 1967. A statistical seismogram model and the filtration problem. *Izv. Acad. Sci. USSR, Phys. Solid Earth*, 6: 356–359 (Engl. ed.).

Yanovskiy, A. K., 1968. Criteria of optimum filtration of random processes in connection with problems of seismic prospecting. *Izv. Acad. Sci. USSR, Phys. Solid Earth*. 6: 361–366 (Engl. ed.).

Zelei, A., 1971. On the design of numerical filters. *Ann. Geofis.*, 24: 457–474.

Zurflueh, E. G., 1967. Applications of two-dimensional linear wavelength filtering. *Geophysics*, 32: 1015–1035.

Zürn, W., 1974. Detectability of small harmonic signals in digitized records. *J. Geophys. Res.*, 79: 4433–4438.

SUBJECT INDEX

Absolute integrability, unit-impulse response, 6, 57
Actual frequency, 64
– output, see True output
Advance time, 79
Aliased pass band, 144
Aliasing, 18, 59, 63
–, spatial, 135
Amplitude array response, 132, 133, 134, 137, 138
– distorted transmission, 14
– impulse modulation, 16
– response, 7, 12, 37, 50, 51
Analog-digital converter, 15, 16
– signal, 15
Anti-aliasing filter, 19
Anticipation function, 123
– term, 38
Apparent velocity, 131, 134
– wavelength, 134
Array alignment, 137
– design, 130
– steering, 137
– tuning, see Array steering
Attenuation, 88
Autocorrelation function, 74, 100, 101, 106, 107, 125
– relationship, 9

Band-limited input, 18
– -pass Butterworth filter, 97
– -pass filter, 77, 96, 97, 104
– -pass infinitely narrow filter, 102
– -pass velocity filtering, 131, 147
– -rejection filter, 77
Bandwidth, 39, 97, 98
Beamforming, 130
Berlage pulse, 72, 128
Bilinear transformation, 63, 64, 91
Binomial array, 143

Block-diagram representation, 44, 131
– – – see also Filter arrangement and Filter realization
– – –, canonic form, 47, 48
– – –, cascade form, see Block-diagram representation, serial form
– – –, direct form, 45, 46, 47
– – –, parallel form, 45, 46, 48
– – –, serial form, 45, 46, 48, 66, 67, 69, 70, 124
Bounded input, 2, 6, 33
– output, 2, 6
– response, 33
Boxcar function, 83
Butterworth circle, 93
– filter, 87, 88

Cauchy theorem, 25
Causal filter, 2
Causality condition, 2, 111
Center frequency, 97, 98
Chebyshev filter, 92
– –, poles, 94
– polynomial, 92
Chirp filtering, 103, 104
– waveform, 103
Complementary components, 18, 37, 80
– pass band, 144
– strip, 26, 61
Compound array, 136
Conjugate filter, 106
– – see also Matched filter
Constant parameters system, 2
– – – see also Time invariant system
Convolution filter, 35
– – see also Finite-memory filter and Non-recursive filter
– integral, 4, 60, 106
– operation, 4, 5
– process, digital, 4

– –, physical, 4
– summation, 29, 35, 36, 41
Corrective filter, 122
Correlation filter, 99, 105
Cross-correlation function, 101, 102, 103
Cross-power spectrum, 103
Cut-off frequency, 39, 64, 77, 87, 91, 95, 96

D.C. rejection filter, 52, 53
Deconvolution filter, 120
– – see also Inverse filter
Delay-and-sum technique, 130, 131, 138
Delayed spiking filter, 126
– summation, 137
Delay time, 79
Desired output, 3, 109, 125
Digital convolution process, 4
– integration, 31
– signal, 15
– system, 27
– system, approximation, 29
Digitization, 15
Dirac delta function, 3
– – – see also Unit impulse
– – –, digital equivalent, 21
Directionality, 117
Directional selectivity, 135
Distortionless filter, 113
– transmission, 13, 14
Downward-continuation method, 148

Enhancement filter, see Smoothing filter and Zero-lag filter
Equivalent low-pass filter, 97
Error sequence, 60, 125
Exponential function, digital equivalent, 21

Fan filtering, 145
– -pass filtering, 145
Feedback filter, 35
– – see also Infinite-memory filter and Recursive filter
– path, 46, 48, 53, 68
– system, 2, 41
Fictitious frequency, 64, 65, 91
– – see also Pseudo frequency
Filter, 1
– analysis, 1
– -and-sum technique, 130, 131
– arrangement, 45

– – see also Block-diagram representation and Filter realization
– realization, 44, 73
– – see also Block-diagram representation and Filter arrangement
– –, canonic form, 75, 76
– –, direct form, 44, 45, 73, 75, 76
– synthesis, 1, 36
Filtre en éventail, 145
Finite length output, 30
– -memory filter, 35, 37
– – – see also Convolution filter and Non-recursive filter
Fixed parameters system, see Constant parameters system and Time invariant system
Folding, 32
– angular frequency, 18
– – – see also Nyquist angular frequency
– frequency, 19, 50, 52, 53
– – see also Nyquist frequency
Forward difference operator, 60
– path, 46, 48, 53, 68
Fourier transform, 7
Frequency dependent attenuation, 13
– – phase shift, 13
– discrimination, 1
– filter, 1
– response function, 7
– – – see also Response function
– -shifting theorem, 17, 97
– -transformation method, 64

Gain factor, 7, 52, 54
Gibbs phenomenon, 39, 81
Gregory formula, 60, 62

Hamming, 40
– window, 40
– –, two-dimensional, 116
Hanning, 40
– filtering, 40
– window, 40
High-pass filter, 77, 94
– – –, Butterworth, 95, 96
– – velocity filtering, 131, 146
Hilbert transform, 8
Horizontal array, 130
– wavelength, 131, 134
– wavenumber, 132, 134

SUBJECT INDEX

Ideal band-pass filter, 98
– – – velocity-transfer function, 145
– filter, 77
– high-pass velocity-transfer function, 145
– – – zero-phase filter, 94
– low-pass filter, 78, 83
– – – phase distortionless filter, 79
– output, 109
Impulse invariance, 55, 61
Infinite-memory filter, 35
– – – see also Feedback filter and Recursive filter
Input autocorrelation, 10, 11
– -output cross-correlation function, 11
– – relation, ideal sampler, 17
– – relationship, analog signals, 4
– – –, digital signals, 29
– – –, first-order recursive filter, 42
– – –, frequency domain, 7
– – –, second-order recursive filter, 44
– power spectrum, 12
Integral operator, 4
Inverse filter, 20, 121
– – see also Deconvolution filter
– –, delayed, 126
– –, least-squares approximate, 124

Kalman filter, 112
k-plane response, 151

Lag filter, 109
– window, see Time window
Laplace transform, 8, 22, 23
– –, bilateral, see Laplace transform, two-sided
– –, one-sided, 8
– –, two-sided, 8
– –, unilateral, see Laplace transform, one-sided
Leakage, 40
Linear array, 134, 143
– – see also One-dimensional array
– filter, 2
– operator, 2
Log-modulus plot, 65
Low-pass filter, 19, 38, 77, 97
– – –, Butterworth, 87, 88, 93
– – –, Chebyshev, 92, 93

Main lobe, 39
Mapping function, 26, 34, 61, 63
Martin-Graham filter, 37, 82, 85, 86, 87

Matched filter, 99, 105, 106
– – see also Conjugate filter
Maximum-delay response, 112
– -likelihood method, 151
Memory terms, 38
Minimum-delay response, 112
Mixed-delay response, 124
Movable strip, 31
Move-out velocity, see Apparent velocity
Multi-channel filter, 2
– – Wiener filtering, 151, 152
Multidimensional filter, 116, 130
Multiple-input system, 2, 118
– – -multiple-output system, 2
– -output system, 2
Multiplication of polynomials, 32

Noise, 1
–, coherent, 143
–, incoherent, 143
–, microseismic, 101, 143
–, nonlinearly polarized, 117
–, periodic, 101
–, random, 100, 103
–, signal-generated, 143
Noncausal time series, 25
Nonrecursive filter, 35, 37
– – see also Convolution filter and Finite-memory filter
– –, high-pass, 95
– filtering, 35, 42
– – see also Transversal filtering
Normal equation, 126
Normalized dipole, 122
Notch filter, 54, 77
– – see also Band-rejection filter
Nyquist angular frequency, 18
– – – see also Folding angular frequency
– criterion, 33
– frequency, 19
– – see also Folding frequency

One-dimensional array, 143
– – – see also Linear array
Operator, 1
–, finite length low-pass, 82
–, optimum, 115
–, prediction, 115
–, symmetric least-squares, 82
–, two-dimensional, 147, 148

Optimized Chebyshev array, 141
Optimum filter, 99, 106
— —, downward-continuation, 149
— —, mean-square error, 111
— —, second-derivative, 149
— —, two-dimensional, 116
— —, Wiener, 99, 108, 134
— system function, 111
Ormsby filter, 37, 82, 84, 87
— —, low-pass, 84
Output "age", 6
— -energy filter, 108
— -invariance technique, 59, 61
— power spectrum, 11
—, sampler, 16
—, time varying linear system, 6

Parabolic approximation, 60, 62
— smoothing, 37
Parametric detector, 105
Partial-fraction expansion, 45, 48, 56, 58, 90
Particle-motion pattern, 117
— — — see also Polarization
Pass filter, 78
Phase response, 7, 12, 50, 51
— —, recursive filter, 65
— -distorted transmission, 14, 38
— -distortionless filtering, 14, 65, 66, 69, 71, 79, 82, 87, 112, 146
— shift, 131, 136
Physically realizable system, 2, 106
Physical system, 1
Pie-slice processing, 145
Polarization, 117, 118
— see also Particle-motion pattern
— discrimination, 1
—, elliptical, 117
— filter, 1, 99, 116, 117, 118
—, rectilinear, 117
Pole-zero technique, 50, 98
Poles, complex-conjugate, 58
—, first-order, 56
—, multiple, 58
Power spectrum, 74, 107
p-plane, 9, 26, 33, 63, 89, 94
Prediction filter, 109
Predictive decomposition, 115
Prewarping, 64, 91
Primary components, 18, 37, 80
— wavelet, 127

Proportionality law, 2
Pseudo frequency, 64
— — see also Fictitious frequency
Pulse-transfer function, 31
— — — see also System function
— — —, Butterworth filter, 90
— — —, optimum filter, 114

Quantization, 15, 71
— effects, 35, 71, 75
— errors, 71, 72, 73, 75
—, filter parameters, 76
— interval, 71, 74, 75, 76
Quantizer, 16

Ramp pulse, 27
Random input, 9
— output, 9
— —, mean-square value, 10, 11
Rectangular approximation, 60, 61, 62
Rectilinearity, 117, 118
Recursive filter, 35, 40, 41, 42
— — see also Feedback filter and Infinite-memory filter
— —, first-order, 42, 72
— —, Lth-order, 46, 73
— —, second-order, 44
Response function, 7
— — see also Frequency response function
— —, frequency-wavenumber, 150
— —, ideal, 77
— —, two-dimensional, 147
— length, 4
Reverse-time filter, 68
Ricker wavelet, 128
Riemann surface, 26, 63
Roll-off, 82, 83, 85, 87
— —, Chebyshev filter, 93
— —, cosine, 86
— —, first-order, 84
— —, linear, 83
Round-off errors, 72, 73, 74
Routh test, 33

Sampled-data system, 28
Sampler, 16
— output spectrum, 17
Sampling, 15
— angular frequency, 17
— function, 16

SUBJECT INDEX 167

— period, 15
— theorem, 19
Second-derivative method, 148
Seismic array, 130
Shallow seismic exploration, 133
Shaping filter, 127
Side lobes, 39, 40
— —, amplitude, 138
Signal, 1
— move-out, 134
—, periodic, 100, 102
— restoration, 5
— step-out, see Signal move-out
—, teleseismic, 143
Single-channel filter, 2
Smoothing filter, 110
— — see also Zero-lag filter
Source wavelet, 127
Spectral window, 39
Stability, 6
— criterion following from Laplace transform relationships, 9, 57
—, first-order recursive filter, 42, 43
—, nonrecursive filter, 36
— region, 34, 51
—, rejection filter, 52
— test, 33
Stable system, 2
— poles, 34, 91
Stationary random signal, 9
Steady-state signal, 12
Straight summation technique, 130, 131, 136
Strakhof's filter, 82
Subarray, 136
Subfilter, 76
Subsystem, 45, 46
Superposition integral, 4
— law, 2
Symmetric filter, 38, 79, 147
— —, band-pass, 97
Symmetry property, response function, 7
— —, pole distribution, 90
System function, 31, 35
— — see also Pulse-transfer function
— —, Butterworth filter, 91
— —, digital filter, 49
— —, first-order recursive filter, 43
— —, nonrecursive filter, 38, 39
— —, polynomial form, 43, 49
— —, second-order recursive filter, 44

— —, zero-phase, 67
— identification, 5

Tangent filter, 92
— —, Chebyshev low-pass, 94
Tapered array, 138
Tapering, 40
— see also Truncation
Terminal relations, 1
Time-domain synthesis, 4
— invariance, 2
— -invariant system, 2
— — — see also Constant parameters system
— — physically realizable system, 6
— -reverse input, 67, 68
— shifting, 137
— -space filtering, 150
— -variable weighting, 6
— -varying deconvolution, 128
— — filter, 2, 112
— window, 39, 97
Transfer function, 8
— —, ideal, 82, 84
— —, optimum three-dimensional, 116
— —, optimum two-dimensional, 116
— —, two-dimensional, 144, 145
Transversal filtering, 35
— — see also Nonrecursive filtering
Trapezoidal approximation, 60, 61, 62
Trigonometric filter, 92
True output, 109, 125
Truncated approximate inverses, 123
— unit-impulse response, 29, 37, 39, 81
Truncation, 40, 81, 97
— see also Tapering
Two-dimensional array, 149
— — wavenumber filtering, 148
— — — — see also Wavenumber-wavenumber filtering

Uniform ripple, 92
Unit impulse, 3
— — see also Dirac delta function
— — response, 3
— — —, band-pass filter, 97, 98
— — —, high-pass filter, 94, 95
— — —, infinitely long inverse filter, 121
— — —, low-pass filter, 78, 79, 80
— — —, Martin-Graham filter, 85
— — —, matched filter, 105

– – –, Ormsby low-pass filter, 84
– – –, time-space, 145, 150
– – –, Wiener optimum filter, 111
– -step function, digital equivalent, 21
– – response, 5

Vector wavenumber, 116, 150
– – and frequency filter, 116
– – filter, 116
Velocity aliasing, 137
– discrimination, 1
– filter, 1
– filtering, 130, 142
Vertical array, 152

Wavenumber, 116
– aliasing, 135
– and frequency filter, 116
– resolution, 135
– -wavenumber filtering, 130, 148
– – – see also Two-dimensional wavenumber filtering

Weighted coefficients, 39, 80
– delay-and-sum technique, 130, 131, 138
– sum, 5
Weighting coefficients, 138, 141
– function, 5, 39, 81, 82, 84
–, two-dimensional, 6
Wiener-Hopf equation, 111, 114
Window carpentry, 39

Zero-lag filter, 110
– – – see also Smoothing filter
– -phase response, 13, 38, 66, 69, 79
z-plane, 26, 33, 34, 63
z^{-1}-plane, 33, 34, 63
z-transform, 20, 22, 23, 24, 34
– –, engineering definition, 24, 26
– –, Laplace definition, 24
– –, one-sided, 24
– –, two-sided, 24, 25

DATE DUE

45-230 Printed in USA